SAFE WORK PRACTICES for the ENVIRONMENTAL LABORATORY

Frank R. Spellman, Ph.D.
Hampton Roads Sanitation District
Virginia Beach, Virginia

TECHNOMIC PUBLISHING CO., INC.
LANCASTER • BASEL

HOW TO ORDER THIS BOOK

BY PHONE: 800-233-9936 or 717-291-5609, 8AM–5PM Eastern Time

BY FAX: 717-295-4538

BY MAIL: Order Department
Technomic Publishing Company, Inc.
851 New Holland Avenue, Box 3535
Lancaster, PA 17604, U.S.A.

BY CREDIT CARD: American Express, VISA, MasterCard

BY WWW SITE: http://www.techpub.com

PERMISSION TO PHOTOCOPY–POLICY STATEMENT

For
Laboratory Technicians Everywhere

Safe Work Practices for the Environmental Laboratory
a **TECHNOMIC**® publication

Published in the Western Hemisphere by
Technomic Publishing Company, Inc.
851 New Holland Avenue, Box 3535
Lancaster, Pennsylvania 17604 U.S.A.

Distributed in the Rest of the World by
Technomic Publishing AG
Missionsstrasse 44
CH-4055 Basel, Switzerland

Printed in the United States of America
10 9 8 7 6 5 4 3 2 1

Main entry under title:
Safe Work Practices for the Environmental Laboratory

A Technomic Publishing Company book
Bibliography: p.
Includes index p. 157

Library of Congress Catalog Card No. 97-61613
ISBN No. 1-56676-574-9

Table of Contents

Preface

THE primary purpose of *Safe Work Practices for the Environmental Laboratory* is twofold: (1) for the person designated as the laboratory's Chemical Hygiene Officer or Safety Officer this text is a user-friendly reference that will provide a format, a template, a guide to compliance with OSHA's Laboratory Standard (29 CFR 1910.1450); and (2) for the person who is assigned to work in the environmental laboratory, this user-friendly text provides the information needed not only to perform routine laboratory tasks correctly but also to perform them safely.

Why concentrate on safety in the environmental laboratory? Good question! Maybe some historical perspective, concerning laboratories, will help to answer this question.

In the not too distant past, when one mentioned the term *laboratory*, it was not unusual for the listener to associate this term with labs he or she was familiar with. Actually, we are talking about the labs most people are familiar with: the university science, research, or medical lab; the high school science lab; and the hospital lab.

When one mentions the term *laboratory* today, a different image or set of images might be generated because, in addition to the types of labs already mentioned, there are other types of labs. For example, in one instance one might form an image of an illicit drug lab—television and films have educated millions of folks on this type of illegal operation (this text does not address this type of laboratory; this makes sense when you consider that this text deals with workers' safety and not with their death and destruction). Because of the illegal products produced in these illegal labs another person might form an image of another laboratory type that has come into existence in recent years: the laboratory that conducts random urinalysis and pre-employment urinalysis for prospective employees of various companies (including the United States Department of Transportation—to certify drivers drug-free for CDL licensing).

Then there is the other laboratory type that some folks might think about when the term *laboratory* is mentioned: the *Environmental Laboratory*. It is this lab that this text is all about. More specifically, as the text title makes clear, it is the performance of safe activities or safe work practices in the environmental laboratory that this text focuses on.

What is an environmental laboratory? Another good question!

Answer: Simply stated, the environmental laboratory discussed in this text is involved with performing analytical testing and sampling protocols relating to air, soil, biosolids, sludges, drinking water, wastewater, groundwater, stormwater, waste characterization, petroleum products, and HRSD/NPDES effluent studies. It should also be pointed out that many wastewater treatment plants and water works have their own environmental laboratories. These labs primarily perform analysis of process conditions to ensure optimization of the process. However, even these small labs (a few are quite large) perform "environmental sampling" and therefore are environmental laboratories.

From the preceding list it should be obvious that the actual genesis of the environmental laboratory can be attributed to the environmental regulations that have been generated by USEPA, AOAC, ASTM, NIOSH, OSHA, and other regulatory and advisory entities.

The typical environmental laboratory contains several different types of hazards the lab worker must guard against. This is the case even though modern environmental laboratories have been designed to take maximum advantage of engineering controls that work to "engineer-out" most hazards.

The main hazard discussed in this text has to do with hazardous materials—dangerous chemicals and compounds—and the affect they can have on work practices. OSHA is quite specific in regards to protecting the laboratory worker from harm that could result from handling hazardous materials—these specifics are discussed in detail throughout this text.

It is important to point out that this text will provide the user with more than just a "safety book." For example, this text provides the user with a sample Chemical Hygiene Plan, it discusses various safe work practices for standard operating procedures normally performed in the environmental laboratory, and it discusses procedures to use for emergency response activities, such as cleanup of chemical spills.

The bottom-line is that probably the most important benefit to be derived from using this text is the exposure the user receives to the lessons learned and examples presented throughout the text; these lessons learned and examples provide information on how to make your environmental laboratory and the performance of your individual work practices safer.

When you get right down to it, isn't this what a safety text should be all about?

Acknowledgments

I would like to thank Mr. Donnie Wheeler, General Manager, Hampton Roads Sanitation District (HRSD), Virginia Beach, Virginia for permission to review and reference many materials belonging to HRSD, used extensively in the writing of this text.

The photographs shown throughout this text came from HRSD's Environmental Laboratory, a state-of-the-art facility operated by artful *experts*.

In having come to the realization that the writing of a text like this one relies on the expertise of many people to "get it right," it has also become apparent to me that there is always one person who contributes effort and expertise above and beyond that requested in making sure that I "get it right." I have been fortunate to have had the advice, critical review, and analysis of such a person—who is also a soulmate—and a genius in those things that are important. For her help, advice, guidance, constructive criticism, friendship, and genius, I thank Joanne Drinan.

FRANK R. SPELLMAN
Virginia Beach, Virginia

PART I

MANAGING HEALTH AND SAFETY PROGRAMS

CHAPTER 1

Introduction

The health and safety of personnel who work in a laboratory is ultimately the responsibility of the person in charge of the facility.

WHILE it is truc that there are some individuals who might take issue with the above statement, the fact is that this statement is true. Moreover, now that the Occupational Safety and Health Administration (OSHA) has implemented its **performance standard**: 29 CFR 1910.1450 *Occupational Exposure to Hazardous Chemicals in the Laboratory* (Laboratory Standard) 1992, not only is the opening statement true it is also, in the United States of America, the "law of the laboratory."

The problem that some laboratory managers or managers of facilities with laboratories have with understanding and accepting their responsibility for the health and safety of their laboratory workers stems from a mindset that alleges that **safety is relative**. These confused managers have nurtured a mindset that has warped their thinking to the point where they are convinced that safety, because they classify it as relative, is not absolute. That is, these individuals feel that safety is dependent or interconnected with something else for intelligibility or significance.

If the reader has difficulty (or is bewildered) in understanding the exact meaning of the above paragraph, he/she joins a long list of those who share a much different meaning of responsibility for safety of workers in the workplace. What the preceding paragraph really conveys is an **attitude**. Since, attitudes, once established, are very difficult to change, the attitude that safety is relative (a way of saying that safety is of secondary importance) has been carried over into many careers, including a significant fraction of current laboratory personnel. This makes sense when one takes into consideration the old adage which states: Employees reflect the attitude of the leader.

The particular instance just related, where certain attitudes held by certain personnel and especially leaders have relegated safety to a level of secondary importance, is an unfortunate situation. This is particularly true when one considers that **the** primary moral responsibility of any manager/leader is to protect the health and safety of his or her workers.

Along with attitudinal problems that often interfere with or prevent installation of viable safety programs into an organization, there has been a recent mindset out there in the "real world" that professes that there is a difference between managers and leaders. Moreover, today everything is either ***pro*** this or ***pro*** that. Pro-active this or that is a common way of stating a point to impress certain people.

What does all this mean?

Does it mean, for example, that if an organization is pro-active in safety (where the goal is the pursuit of zero accidents or mishaps in the workplace) it is practicing the kind of pro-active activity that makes sense? Or, on the other hand, if an organization states that it is pro-active in safety and also states that its managers are different from its leaders, then is there a problem? (Proponents of this particular philosophy feel that managers are different than leaders because managers make the rules and leaders enforce them).

The answer to both of these questions can be stated quite simply: Yes; certainly!

This text takes the view that without top management support of the organization's safety program, there is **no** program. Management, from the top to the bottom level, that views safety as relative (and therefore of secondary importance) is best characterized as dysfunctional.

Additionally, while it is true that managers make the rules, this text takes the view that managers are also the leaders that enforce the rules—Managers **cannot** delegate safety; in the end, they (Managers) are ultimately responsible. When you think about it, is this not the trend demonstrated by recent court decisions?

The point is that as leaders, these managers must be able to motivate their workers toward a certain attitude and behavior that does not describe safety as being relative or as being of secondary importance.

If the reader feels as the author does that a viable organizational safety program is paramount in regards to maintaining the good health and safety of all workers, then this text is for you. Moreover, if the reader shares a **pro-active** view of this topic, then this is certainly in line with the view taken by certain regulatory enforcement agencies; namely, OSHA.

1.1 SCOPE OF TEXT

Safe Work Practices for Environmental Laboratories is specifically written and organized to address the management, identification, and control of health and safety hazards in the environmental laboratory.

There are many texts that provide thorough coverage of laboratory safety. A big need, however, always exists for a text that does so in a readable format, since comprehensive coverage does not necessarily mean *comprehensible* coverage. Simply stated, *Safe Work Practices for Environmental Laboratories* is a user friendly book.

The reader's immediate question might be: Why the focus on the *environmental* laboratory?

This is a legitimate question.

Some might say that a laboratory is a laboratory; thus, laboratories are all the same. However, in this instance, nothing could be further from the truth.

It is true that in earlier times, approximately 20 years ago, when one thought about laboratories he/she usually thought about laboratories limited to traditional settings such as in hospitals, universities, research facilities and in chemical industries. However, because times have changed, laboratories are now commonplace. This change was brought about primarily by the growth of environmental regulations which have, in turn, caused a corresponding growth in the number of laboratories that are capable of performing analysis of environmental samples. This increase in environmental laboratory activity has also increased the need for health and safety programs in the environmental laboratory.

To gain a better understanding of how profound this growth in the number of environmental laboratories has been, one need only review the escalation of environmental regulations that has occurred in recent years. Such environmental regulations that deal with clean air, clean water, cleanup of waste disposal sites, handling hazardous materials, regulations dealing with the handling and disposal of biohazards, and nuclear regulations have combined not only to create an economic opportunity for entrepreneurs who would own laboratories, but more importantly, also have worked to drive up the need for additional environmental laboratory workers.

This text also focuses on the environmental laboratory for another reason. There are several excellent texts available on laboratory safety that target all laboratories in general but few that focus on the hazards that are inherent with the environmental laboratory. For example, the traditional environmental laboratory in water and wastewater treatment facilities is different from laboratories in hospitals, universities, and research facilities. While it is true that laboratories located in any of these facilities have many of the same types of hazards as any other laboratory, it is also true that there are some profound differences. As a case in point, consider the wastewater treatment plant laboratory. The typical wastewater treatment laboratory is suited for the analysis of specific waterborne substances where their primary function is to provide support for process-control and permit-compliance required tests. They do not analyze the types of substances or perform the types of tests that hospitals, universities, and research laboratories do. The wastewater laboratory is different.

Further, the water or wastewater laboratory worker is exposed to chemical hazards that are not common in other types of laboratories. These chemical hazards might include chlorine, sulfur dioxide, anhydrous ammonia, ferric chloride, arsenic, cyanide, alum, muriatic acid, sodium hydroxide, and others.

The point is that laboratory activities vary widely from laboratory to laboratory. For example, the hospital laboratory worker has to guard against contamination from infected blood and/or tissue samples and radiation exposure that most environmental lab workers are not exposed to. At the same time, the university lab worker may be exposed to all types of known and unknown hazards, while the worker in the environmental lab normally would be exposed to a narrower range of hazards. In the case of the water/wastewater lab worker, exposure is limited to an even narrower range of hazards specific to that industry.

This is not to say that the lab worker in all water/wastewater and/or other types of environmental laboratories cannot or will not be exposed to a wide range of hazards. This is certainly not the case for the lab worker in a typical analytical environmental laboratory whose business it is to sample all types of industrial waste, by-products, routine process control samples, etc., these types of samples obviously present the lab worker with a wide-range of hazards. The point is that this text does not specifically address hazards associated with medicinal wastes and exposure to nuclear radiation and other hazards that are usually addressed in general texts that describe *the* laboratory in a broader sense.

The focus here is narrow, to the point, and specifically speaks to a large number of environmental laboratories that range from very small size (closet-sized) to the large

TABLE 1.1. Environmental/Analytical Testing.

(1) Water/wastewater/stormwater/groundwater
(2) Soils
(3) Hazardous waste characterization
(4) Drinking water certification
(5) State NPDES
(6) Petroleum products testing
(7) Field sampling
(8) Lead and paint testing
(9) Asbestos testing

centralized environmental laboratories that are traditionally part of, for example, a large sanitation district (see Figures 1.1, & 1.2).

The environmental laboratory that this text is designed to **target** could expose the laboratory worker to hazards that are inherent with the type/kinds of analytical testing listed in the Table 1.1.

1.2 OBJECTIVES

The main objective of this text is to stress the two major components of laboratory safety: (1) the design of safe facilities and (2) safe work practices. It should be pointed out that no matter how well the laboratory is designed, improper usage of laboratory facilities (improper work practices) can always defeat the safety features designed into the lab, or, for that matter, anywhere else.

Following the guidelines provided in OSHA's Laboratory Standard and OSHA's *Guidelines on Workplace Safety and Health Program Management* (1989), another objective of this text is to point out the four main elements that must be woven throughout the fabric that makes up any viable laboratory safety program. These elements include the following:

(1) Management leadership and employee involvement
(2) Work site analysis
(3) Hazard prevention and control
(4) Safety and health training

In line with its objectives, this text is composed of seven parts and 16 chapters that cover the wide range of topics that are important in maintaining a safe laboratory environment. Moreover, the main elements that make up OSHA's Laboratory Standard are specifically covered in various chapters. For example, the management leadership, employee involvement, and administrative requirements are discussed in Chapter 2. Work site analysis (hazard evaluation and identification) is addressed in Chapter 3. Safety and health training is addressed in Chapter 5. Hazard prevention and control procedures, the <u>core</u> of any safety program, are discussed throughout this text.

1.3 TERMINOLOGY

Before we discuss specific safety elements or topics covered, it will be beneficial for the reader to review the definitions of the key terms used in this text. Normally, key terms like the ones that follow are listed in an end-of-text glossary. Experience has shown that glossaries are beneficial to the reader. However, definitions of key terms that are provided early in the text, to set the stage for the material that follows, enables the reader to gain greater understanding of the subject matter that follows.

1.3.1 DEFINITIONS OF KEY TERMS

Accident. An event occurring by chance or from unknown causes (and therefore unpreventable) (Grimaldi & Simonds, 1989).

Action Level. The exposure level (concentration of the material in air at which OSHA regulations to protect employees take effect); for example, workplace air analysis, employee training, medical monitoring, and record keeping. Exposure at or above the action level is termed occupational exposure. Exposure below this level can also be harmful.

Acute Effect. An adverse effect on a human or animal, characterized by severe symptoms that develop rapidly.

Acute Toxicity. The adverse effects resulting from a single (or small number) dose or exposure to a substance.

ANSI (American National Standards Institute). A privately funded organization that identifies industrial and public national consensus standards and oversees their development. Many of these standards relate to safe design and performance of equipment and safe work practices and procedures.

Autoignition Temperature. The minimum temperature at which a substance ignites without application of a flame or spark.

Blood-borne Pathogen. According to OSHA, pathogenic microorganisms that are present in human blood and in other potentially infectious materials and can cause disease in humans (e.g., HIV and hepatitis B).

Boiling Point. The temperature of a liquid at which its vapor pressure is equal to or very slightly greater than

Figure 1.1. Top photo shows front view of an environmental laboratory; bottom photo shows rear side view of same building.

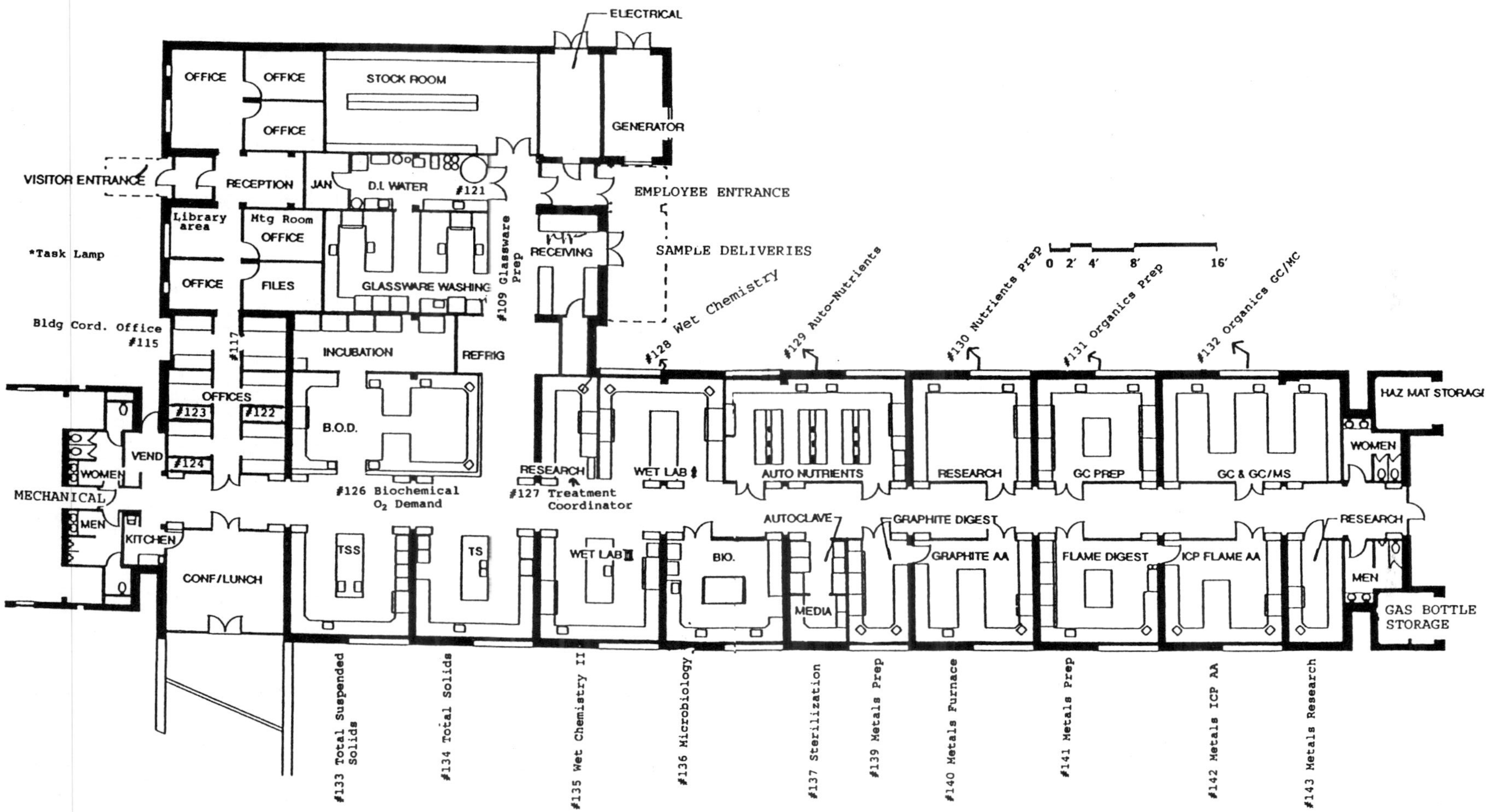

Figure 1.2. Shows floor plan of environmental laboratory shown in Figure 1.1.

the atmospheric pressure of the environment. Flammable materials with low boiling points generally present special fire hazards.

CERCLA. The Comprehensive Environmental Response, Compensations, and Liability Act. The Superfund Law PL 96-510, found at 40 CFR 300. The EPA has jurisdiction. Enacted December 11, 1980, and amended thereafter, CERCLA provides for the identification and the cleanup of hazardous materials that have been released over the land into the air, waterways and groundwater. It covers areas affected by newly released materials and older leaking or abandoned dump sites where hazardous materials have been released. The EPA has authority to collect the cleanup costs from those who release the waste material. Cleanup funds come from fines and penalties, from taxes on chemical/petrochemical feed stocks, and the U.S. Treasury Department. A separate fund collects taxes on active disposal sites to finance their monitoring after they are closed. CERCLA is a result of the serious problems that arose from the release of hazardous materials at the Love Canal near Niagara Falls, NY, in August 1978.

CFR (Code of Federal Regulations). A compilation of all current regulations and standards published by the Office of theFederal Register.

Chemical Hygiene Officer (CHO). As per 29 CFR 1910.1450, the CHO is an employee who is designated by the employer and who is qualified by training and experience to provide technical guidance in the development and implementation of the provisions of the Chemical Hygiene Plan. This definition is not intended to place limitations on the position description or job classification that the designated individual shall hold within the employer's organization structure.

Chemical Hygiene Plan (CHP). A written plan that specifies the procedures the employer has implemented to ensure that employees are protected from exposures to hazardous chemicals. The plan should address such elements as work practices, controls, protective equipment, engineering controls, training, standard operating procedures, and medical evaluations.

Chemical Reactivity. The ability of a material to chemically change, possibly resulting in hazardous effects such as heat, explosions, or the production of toxic substances.

CHEMTREC (Chemical Transportation Emergency Center). A national center established by the Chemical Manufacturers Association (CMA) in Washington, DC, in 1970, to relay pertinent emergency information concerning specific chemicals on request. CHEMTREC has a 24-hr toll-free telephone number (1-800-424-9300), intended primarily for use by those who respond to chemical transportation emergencies.

Chronic Health Effect. An adverse effect on a human or animal with symptoms that develop slowly over a relatively prolonged period of time.

Chronic Toxicity. Adverse effects resulting from repeated doses or exposures to a substance over a relatively prolonged period of time.

Clean Air Act (CAA). Public Law Pl 91-604, with regulations enacted by the EPA found at 40 CFR 50-80. The regulatory vehicle authorizing the EPA to control and monitor airborne pollution hazardous to public health or natural resources.

Clean Water Act (CWA). Public Law Pl 92-500, with regulations enacted by the EPA found at 40 CFR 100-140 and 400-470. The CWA regulates the discharges of nontoxic and toxic pollutants into surface waters.

Combustible Liquids. Any liquid having a flashpoint at or above 100°F but below 200°F, except any mixture having components with flashpoints of 200°F, or higher, the total volume of which make up 99 percent or more of the total volume of the mixture. Combustible liquids do not ignite as easily as flammable liquids but can be ignited under certain circumstances and must be handled with caution.

Compressed Gas

- a gas, or mixture of gases, having, in a container, an absolute pressure that exceeds 40 psi at 70°F
- a gas, or a mixture of gases, having, in a container, an absolute pressure that exceeds 104 psi at 130°F,regardless of the pressure at 70°F
- a liquid having a vapor pressure that exceeds 40 psi at 100°F

Concentration. The relative amount of a substance when combined or mixed with other substances. Often expressed in mg/m^3, mg/kg, or ppm.

Containment (biological). Safe methods for managing infectious agents in the laboratory environment where they are being handled or maintained.

Department of Labor (DOL). U.S. Department of Labor; includes the Occupational Safety and Health Administration (OSHA).

Department of Transportation (DOT). Regulates transportation of materials to protect the public as well as fire, law, and other emergency-response personnel.

Dermal Toxicity. Adverse effects resulting from skin exposure to a substance. Effects may be local or result from absorption through the skin (systemic).

Emergency. Any occurrence, such as, but not limited to rupture of containers, equipment failures, or failure of control equipment that results in an uncontrolled release of hazardous chemical into the workplace.

Engineering Controls. Systems that reduce potential hazards by isolating the worker from the hazard or by removing the hazard from the work area. Engineering controls are preferred over personal protective equipment (PPE) and should be used in conjunction with administrative practices.

Explosive Limits. Upper explosive limit (UEL) and lower explosive limit (LEL).

Flammable Gas

- a gas that, at ambient temperature and pressure, forms a flammable mixture with air at a concentration of 13 percent by volume or less
- a gas that, at ambient temperatures and pressure, forms a range of flammable mixtures with air wider than 12 percent by volume, regardless of the lower limit

Flammable Limits. The minimum and maximum concentrations of a flammable gas or vapor between which ignition can occur.

Flammable Liquid. Any liquid having a flashpoint below 100°F, except any mixture having components with flashpoints of 100°F or higher, the total of which makes up 99 percent or more of the total volume of the mixture.

Flammable Solid. A solid, other than an explosive or blasting agent, that is likely to cause fire through friction, absorption of moisture, spontaneous chemical changes, or retained heat from manufacturing or processing, or that can be ignited readily and, when ignited, burns as vigorously and persistently as to create a serious hazard.

Flash Point. The minimum temperature at which a liquid gives off vapor within a test vessel in sufficient concentration to form an ignitable mixture with air near the surface of the liquid.

Fume. Airborne particulate formed by the vaporization of solid materials; usually refers to metals. Fume particles are usually less than 1 micron (μm) in diameter.

Fume Hood. Part of local exhaust system in which airborne fumes are drawn.

General Ventilation. System of ventilation consisting of either natural or mechanically induced fresh air movements to mix with and dilute contaminants in the workplace air.

Hazard. Biological, chemical, or physical conditions that have the potential for causing harm to people, property, or the environment.

Hazardous Chemical. According to OSHA, a hazardous chemical is a chemical for which there is statistically significant evidence based on at least one study conducted in accordance with established scientific principles that acute or chronic health effects may occur in exposed employees.

Hazardous Material. According to the DOT, a hazardous material is any substance or compound that has the capability of producing adverse effects on the health and safety of humans.

Hazardous Substance. According to CERCLA, any substance that is designated as a hazardous substance under the Clean Water Act and the Clean Air Act or any substance designated as a hazardous waste under RCRA.

Health Hazard. Includes chemicals that are carcinogenic, toxic or highly toxic agents, reproductive toxins, irritants, corrosives, sensitizers, as well as agents that damage the lungs, skin, eyes, or mucous membranes.

Ignition Temperature. The lowest temperature at which a combustible material ignites in air and continues to burn independently of the source of heat.

Ingestion. The taking in of a substance through the mouth.

Inhalation. The breathing in of a substance in the form of a gas, vapor, fume, mist, or dust.

Laboratory. As defined by OSHA in 29 CFR 1910.1450, a facility where the laboratory use of hazardous chemicals occurs, where relatively small quantities of hazardous chemicals are used on a nonproduction basis.

Laboratory Use. According to OSHA in 29 CFR 1910.1450, laboratory use of hazardous chemicals occurs when all of the following conditions are met:

- Chemical manipulations are carried out on a laboratory scale.
- Multiple chemical procedures or chemicals are used.
- Procedures are neither part of, nor support, a production process.
- Protective lab practices and equipment are available and in common use to minimize the potential for employee exposure to hazardous chemicals.

Local Ventilation. The drawing off and replacement of contaminated air directly from its source.

MSDS (Material Safety Data Sheet). Information sheets (required by OSHA in 29 CFR 1910.1200) containing data on a chemical's physical properties, health hazards, toxic effects, fire hazards, and so forth.

NFPA (National Fire Protection Association). An international voluntary membership organization to promote/improve fire protection and prevention and establish safeguards against loss of life and property by fire. Best known for publishing the **National Fire Codes**, standards, and recommended practices.

OSHA. Occupational Safety and Health Administration of the U.S. Department of Labor; federal agency with safety and health regulatory and enforcement authorities for most U.S. industry and business.

Oxidizer. A chemical that initiates or promotes combustion in other materials thereby causing fire either of itself or through the release of oxygen or other gases.

Particulate. Small, separate pieces of airborne material. Dusts, smokes, fumes, mists, and fogs are examples of particulates.

Particularly Hazardous Substance. According to OSHA's 1910.1450, these include select carcinogens, reproductive toxins, and substances that have a high degree of acute toxicity.

PEL (Permissible Exposure Limit). Limits developed by OSHA to indicate the maximum airborne concentration of a contaminant to which an employee may be exposed over the duration specified by the type of PEL assigned to that contaminant. May be expressed as a time-weighted average (TWA) or as a ceiling exposure limit that must never be exceeded instantaneously.

Performance Standard. OHSA standard where there are few specific requirements to carry out certain procedures in a certain way. A performance standard allows the party attempting to comply with a certain amount of "license" in achieving compliance.

pH. A scale from 0 to 14 that represents the acidity or alkalinity of an aqueous solution; 7 on the scale is neutral.

Physical Hazard. In regards to a chemical, there is scientifically valid evidence that it is a combustible liquid, a compressed gas, explosive, an organic peroxide, flammable, an oxidizer, pyrophoric, water reactor or unstable.

PPE (Personal Protective Equipment). Devices or clothing worn to help protect a worker from direct exposure to hazardous materials (e.g., gloves, respirators, goggles, face shields, laboratory coats).

Pyrophoric. A chemical that ignites spontaneously in air at a temperature of 130 degrees F or below.

RCRA (Resource Conservation and Recovery Act). Public Law Pl 94-580, with regulations enacted by the EPA found at 40 CFR 240-271, Federal environmental legislation, administered by EPA, aimed at controlling the generation, treating, storage, transportation, and disposal of hazardous wastes.

Reactivity. Describes a substance's tendency to undergo chemical reaction either by itself or with other materials with the release of energy.

Risk. The probability of incurring harm from a hazard. More specifically, it is the assumed effect of an uncontrolled hazard, appraised in terms of the *probability* it will happen, the maximum *severity* of any injuries or damages, and the public's *sensitivity* to the occurrence (Grimaldi & Simonds, 1989).

Route of Entry. The method by which a material enters the body; includes absorption (eye or skin contact), ingestion, and inhalation.

RQ (Reportable Quantity)

- *CERCLA*: quantity of substance designated under CERCLA as hazardous, and the release of which must be reported to the NRC. RQs can be found in Section 311 of the Clean Air Act.
- *DOT*: quantity specified for a substance in the appendix to the Hazardous Materials Table.
- *SARA*: quantity specified in Title III, Section 304, which requires specific reporting.

SARA (Superfund Amendments and Reauthorization Act). A revision and extension of CERCLA, administered by the EPA, intended to encourage and support local and state emergency planning efforts. Title III of SARA is known as the Emergency Planning and Community Right-to-Know Act.

SPCC. Spill Prevention, Control, and Countermeasure Plan.

Specific Gravity. The weight of a material compared to the weight of an equal volume of water.

TLV (Threshold Limit Value). The time-weighted average (TWA) of airborne concentration of contaminant to which nearly all workers may be exposed, for a normal 8-hr workday and a 40-hr workweek without adverse effect. TLV is a recommended limit while the Personal Exposure Limit (PEL) is the legal limit as set and enforced by OSHA.

Toxicity. The sum of adverse effects resulting from exposure to a material.

TSCA (Toxic Substances Control Act). Public Law PL 94-469, with regulations enacted by the EPA found at 40 CFR 700-799. Federal environmental legislation, administered by EPA, for regulating the manufacture, handling, and use of materials classified as "toxic substances."

Vapor Density. The weight of a vapor or gas compared to the weight of an equal volume of air; an expression of the density of the vapor or gas. Materials lighter than air have vapor densities less than 1.0. Materials heavier than air have vapor densities greater than 1.0.

Water Reactive. A chemical that reacts with water and may release gases that are either flammable or present a health hazard. Usually flammable solids will react in varying degrees with water or humid air.

CHAPTER 2

Management, Leadership, and Employee Involvement

Each employer—
(1) Shall furnish to each of his employees employment and a place of employment which are free from recognized hazards that are causing or likely to cause death or serious physical harm to his employees;
(2) Shall comply with occupational safety and health standards promulgated under the OSH Act.

Each Employee—
(1) Shall comply with occupational safety and health standards and all rules, regulations, and orders issued pursuant to this Act which are applicable to his own actions and conduct. (OSHA, 1970, p.7)

2.1 INTRODUCTION

IN Chapter 1 it was pointed out that the supervisor in-charge of the laboratory is ultimately responsible for safety in the laboratory. However, it usually comes as a total surprise to lab workers (and workers in other industries) when they are informed about the statement above relating to employees' responsibility for safety and that they are required by Federal Law to abide by safety rules.

Further, in Chapter 1 it was also made clear that this text takes the view that a viable safety program in any organization, department, division, and/or work center is dependent upon top management's support. This point is made quite clear in the following statement: *"No matter how many regulations, standards, and laws are designed to ensure worker's safety and no matter how experienced and/or motivated the plant's designated safety official is, he/she is powerless without strong support from the highest levels of . . . management" (Spellman, 1996, p. 7).*

Simply put, without a strong commitment from upper management, the organization's safety effort is doomed.

Unfortunately, not all organizations feel that safety in the workplace requires top management support. This is an unfortunate occurrence (and/or mindset). Sometimes safety is viewed as a necessary evil. Moreover, many managers view safety as a ***cost*** rather than as a benefit. While it is true that implementing a viable safety program is not without financial expense, it is also true that on-the-job injuries (OJIs), lost-time away from work because of OJIs, destruction of equipment, injuries to other employees, mishaps or accidents that cause damage to the environment, and fatalities are also expensive. With the trend today being along the lines of ***let me sue you before you sue me***, does it not make good financial sense to comply with safety and health regulations that are designed to prevent OJIs, provide for a healthier place to work and, in the long-run, save large sums of money?

Earlier it was pointed out that workers have a responsibility for safety in the workplace. This is an important point and the organizational safety official should take into account that workers can be valuable assets in putting together a strong workplace safety program. It is important not only to enlist input from laboratory workers but is also important to state in the lab's Safety Plan/Chemical Hygiene Plan that all workers are responsible for safety in the Environmental Laboratory.

If you are or become the laboratory safety officer, a question you should ask yourself is: How many safety plans, manuals, texts, or other lab documents clearly identify the person who is designated responsible for safety? You have probably read several such documents. The next question becomes: What about those lab workers who are not designated by various laboratory documents as being responsible for safety? Shouldn't everyone share this responsibility? This text takes the view that every person working in the Environmental Laboratory (or any other work center) is responsible for safety. Simply stated, the organizational culture within the Environmental Laboratory be built on and around maintaining workplace safety.

In order to demonstrate top management's view that safety really is number one in the workplace, it is necessary for top management to clearly state the organizational policy on safety and health issues in the workplace. This organizational safety and health policy should be issued as a formal written document. The following section discusses this important element in detail.

2.2 THE ORGANIZATION'S SAFETY POLICY

As stated above, management should propose that an organizational safety policy be written and approved by the **"top official"** in the organization. The importance of such an organizational safety policy cannot be overstated. For example, you would not expect an architect to design or a builder to build a skyscraper without cornerstones. Likewise, no organization, large or small, should fashion an organization's safety program without its major cornerstone: The Organization's Safety Policy.

Not only will a well-written organizational safety policy function as the organization's cornerstone for its workcenter safety programs, it will also set the correct tone for conducting organizational functions in the safest manner possible.

This may sound like a tall order.

It is.

The point is that without the organizational safety policy, safety cannot be properly woven into the fabric that makes up the organizational culture.

Experience has shown that to be effective an organizational safety policy statement must not only be well-written, it must also be short and to the point. Workers will not read documents that are voluminous in nature; workers will not understand documents that are voluminous; and workers will not care about documents that are voluminous. When you get right down to it, who does?

It is important to ensure that workers read and understand the organization's safety policy. There are several examples of safety policies used by Fortune 500 companies and others that can be used to model your own organizational safety policy. Some organizations issue two copies of the organizational safety policy statement to new employees. One copy is for the employee to keep and the other is to be read, signed, and dated by the employee when received; then the signed copy is turned back to the issuing official. This signed and dated copy becomes part of the employee's training record and personnel file.

An example of a short, to the point, and hard-hitting safety policy is provided in Figure 2.1.

2.3 GOALS AND OBJECTIVES

In order for workers in the environmental laboratory to understand what is expected of them, in the realm of safety, health, and safe work practices, it is necessary for them to understand the results desired and the measures planned for achieving them. This can only be accomplished by setting clear goals that are achievable. For example, a goal to *reduce* all hazards in the environmental laboratory is a goal that can be clearly understood by lab workers. Moreover, this is a goal that is realistically achievable. On the other hand, if the environmental lab sets a goal to *eliminate* all hazards in the laboratory, then this is an admirable goal but has little chance of succeeding.

Once a realistic goal has been set, such as the one to ***reduce*** all hazards in the laboratory, then it is necessary to establish steps (objectives) required to achieve this goal. It is important to point out that these objectives should not only be attainable but they should also be specific and measurable. An example could include developing a weekly or monthly audit system. This audit or inspection system should be in writing. It does little good to come up with objectives or steps to take to attain a certain goal unless these objectives are written down and passed out to all members of the audit team.

The second step in setting up an in-house safety and health audit team is to assign team members. These team members may be volunteers or may be designated and assigned by management. However they are assigned to the audit team, these team members should be thoroughly familiar with laboratory operation. Moreover, team members should be familiar with auditing procedures and techniques. One way to ensure that audit team members inspect and look for the right kinds of items (hazards) is to have them follow a guidelist or checklist. This checklist should include a list of the common types of hazards that are found in the environmental laboratory. Table 2.1 lists many of the items that should be included in the laboratory audit checklist.

The checklist items listed in Table 2.1 are standard safety items that should be checked for any type of laboratory. Keep in mind, however, that Table 2.1 does not list every possible hazard that should be checked. Some safety items will vary from lab to lab. The point is that in order to conduct a viable and meaningful audit it is necessary to utilize some type of audit checklist.

When the audit is completed, there should be some mechanism in place so that discrepancies can be reported to responsible personnel. Obviously, if those responsible

NO JOB IS SO CRITICAL AND NO SERVICE

IS SO URGENT—THAT WE CANNOT

TAKE THE TIME TO PERFORM OUR

WORK SAFELY.

While it is true that the major emphasis is on efficient operations, it is also true that this must be accomplished with a minimum of accidents and losses. I cannot overemphasize the importance that the District places on the health and well-being of each and every employee. The District's commitment to occupational health and safety is absolute. HRSD's safety goal is to integrate hazard control into all operations, including compliance with applicable standards. I encourage active leadership, direct participation, and enthusiastic support of the entire organization in supporting *our* safety programs and policies.

____

General Manager

Figure 2.1. Sample of an organizational safety policy statement used by Hampton Roads Sanitation District (HRSD), Virginia Beach, Virginia. HRSD's safety policy is similar to several other safety policies that are currently being used in wastewater and other industries. The major point is made in as short of verse as is necessary; thus, this is the type of safety policy that will be read by the employee. More importantly, HRSD's safety policy sends the desired message. Courtesy of HRSD.

are not made aware of safety and health hazards, then it is unlikely that they will be corrected.

It is important to maintain a file of all previous discrepancies discovered during the lab audit. These discrepancies should always be checked first during subsequent audits. By doing this a followup procedure is established that can go a long way toward correcting discrepancies and for measuring the success of the lab auditing procedure. The long run goal, of course, is to provide a laboratory workplace where hazards have been reduced—thus providing a safer place to work.

2.4 LABORATORY SAFETY OFFICIAL'S RESPONSIBILITY/ AUTHORITY

It is common practice for any organization to spell out in clear terms what a particular employee's responsibility is. It is not always clear, however, what the employee's authority is. In the environmental laboratory it is important that the laboratory safety official/officer be assigned in writing the exact nature and extent of his/her responsibility. In addition, the lab safety official/officer must be given enough authority to ensure actual compliance with applicable safety rules and regulations.

In regards to responsibility, the responsibility for safety and health in any work environment resides with ***everyone*** who works in the laboratory. Having said this, it is also important for the lab manager to understand that he/she can assign anyone to be the person overall in charge of safety in the laboratory. What the lab manager cannot do is assign "away" the ***ultimate responsibility*** for the safety and health of all workers. As OSHA makes perfectly clear in its Laboratory Standard, the Manager of the laboratory is ultimately responsible for safety in the lab.

The question becomes: What else is the laboratory officer or supervisor responsible for in regards to safety in the environmental laboratory? This is the question that

TABLE 2.1. Environmental Laboratory Audit Checklist.

1. Housekeeping satisfactory
2. Aisles clutter free
3. Hazard warning signs posted where required (e.g., at entrances, chemical storage areas, poison lockers, acid or corrosive cabinets)
4. Chemical Hygiene Plan available to all employees
5. Material Safety Data Sheets available to all employees
6. Report of Unsafe Condition(s) Form available to all employees
7. Laboratory refrigerators are not used to store food (lunches)
8. Laboratory equipment is in good operating condition
9. Flammable materials are properly stored in properly marked cabinets
10. Fire extinguishers are strategically placed throughout lab; they are clearly visible and marked; maintenance requirements are clearly posted and up-to-date; each extinguisher is in good operating condition and fully charged
11. Flammable materials are maintained in limited quantities (not over-stocked)
12. Doors in hazard labs are well-separated, doors swing outward
13. Chemicals are stored according to compatibility
14. Chemical waste is properly labeled and stored in designated area
15. All containers are properly labeled according to the Chemical Hygiene Plan and Hazard Communication Standard
16. Chemicals are stored on designated shelves; no chemicals on floor
17. Gas cylinders are properly labeled, strapped firmly in place, cylinders not in use are capped, reducing and oxidizing agents are properly separated
18. All work generating hazardous or toxic fumes is done in hoods
19. Fume hoods have been inspected and certified safe for use at least once each month or quarter
20. The laboratory areas should be maintained at negative pressure with respect to corridors
21. Fume hoods are capable of maintaining at least 100 fpm face velocity with the sash fully open
22. Fume hoods are not used for storage
23. All electrical circuitry is in good material condition
24. No circuits are overloaded with extension cords
25. Electrical cords are protected or placed in raceways
26. Adequate lighting is provided
27. Electrical breaker panel is readily accessible
28. Deluge showers located in such a fashion so as to preclude contact with electrical equipment, components, circuitry, etc
29. GFCIs (Ground Fault Circuit Interrupters) are installed as needed
30. Safety showers available, functional, and inspected as required
31. Eyewash stations available, functional, and inspected as required
32. First aid kits available, fully stocked, and inspected as required
33. Personal protective equipment available, in use, used correctly, and properly maintained—goggles, faceshields, gloves, aprons, respirators, etc.
34. Evacuation routes clearly marked and unobstructed
35. No smoking signs posted

any responsible lab supervisor should ask him/herself. Generally, there are five main areas for which the laboratory supervisor has overall safety responsibility. He/she has responsibility to ensure that:

(1) An appropriate safety orientation has been given to individuals when they are first assigned to the laboratory. It should be pointed out that OSHA, under its 29 CFR 1910.1200—Hazard Communication Standard, requires that employees are not to be exposed to chemical hazards until they have been properly briefed and trained on the hazards. This is an important point and many environmental laboratories include this requirement in their Chemical Hygiene Plans.

(2) Information on special or unusual hazards is provided to lab workers anytime they must perform nonroutine work. This is another important point because in this day and age of lawsuits springing up for just about anything, the last thing any supervisor wants to do is to make him/herself liable for the unwanted actions of others. The point is that whenever a supervisor assigns a worker to perform work that the worker is not qualified to do, the supervisor is perching him/herself on the edge of a very precarious cliff.

When assigning workers to nonroutine work in the laboratory, or any where else for that matter, be sure that you ask the worker if he/she is familiar with the nonroutine work. If the worker says that he/she is not familiar with the work and the hazards involved, then the supervisor is responsible to ensure that the worker is briefed/trained prior to beginning the nonroutine work.

(3) The environmental laboratory supervisor must ensure that his/her workers are provided with readily available and adequate safety equipment that is in proper working order. For example, a worker may be assigned to work with a certain chemical in the laboratory. The worker may or may not be familiar with the chemical. Either way, the worker should always review the chemical's material safety data sheet (MSDS) (see Figure 2.2). From the MSDS, the lab worker can determine the hazards involved with working with the chemical. More importantly, the lab worker can also determine the types of safety equipment that should be used when handling the chemical. For example, the MSDS might call for the use of special goggles, faceshields, gloves, and respirators (see Figures 2.3 and 2.4). The laboratory supervisor must ensure that the lab worker has all the required safety equipment available to him or her when needed.

(4) The environmental laboratory supervisor is responsible to ensure that adequate training has been provided

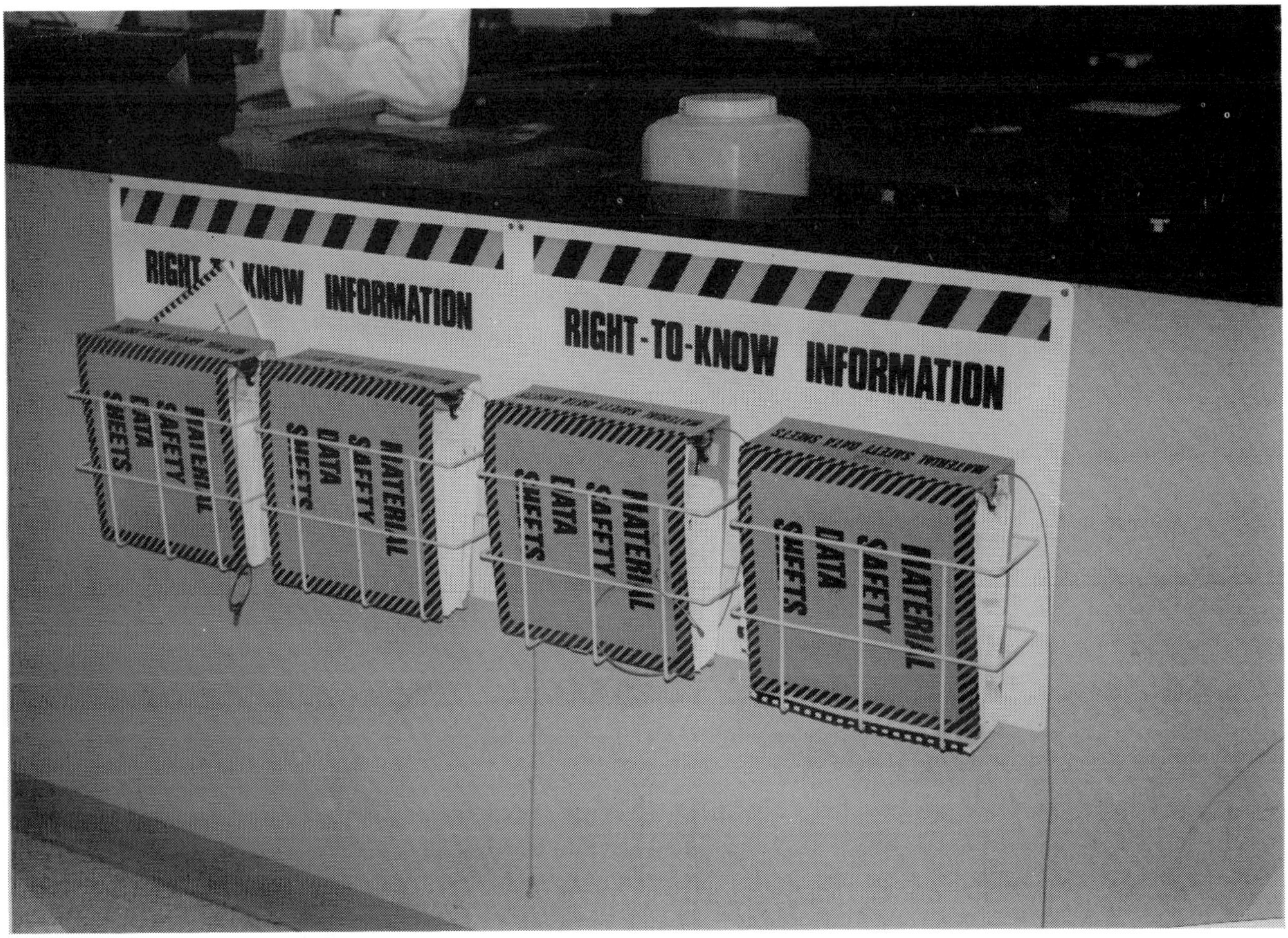

Figure 2.2. Right-to-know station, with MSDS, in an environmental laboratory.

on the proper use of all safety and emergency equipment. This training must be documented. More will be said on the vital topic later.

(5) The laboratory supervisor is responsible to ensure that the lab has a set of published and posted safety rules. In short, safety rules must be visible, must be clearly written, easy to understand, and strictly enforced.

Experience has shown that we are a society engulfed by rules. We have rules for everyone and for every situation. All through our lives we have functioned according to some set of rules. Most people do not like rules. Workers generally do not like rules—who does? The is especially the case when the rules are unclear and/or arbitrary. In putting the environmental laboratory's safety rules together it is wise for the lab supervisor to keep them simple,—so they are easy to understand. Having a set of safety rules is one thing, having effective safety rules is another. One point is certain, you can create all the safety rules you want but if there is no enforcement policy—the rules are worthless. The laboratory supervisor must ensure that safety rules exist and that they are vigorously enforced.

It is wise to make as few rules as possible. Putting together volume after volume of safety rules will serve some dust-collecting shelf well—but they will not serve the purpose for which they were intended: to protect the well-being of all laboratory workers.

In regards to the laboratory safety/officer's authority, the degree and extent of this authority is important. OSHA in its *Guidelines on Workplace Safety and Health Program Management (54 CFR 3908) (1989)* makes it clear that the employer has a responsibility to provide adequate authority and resources to responsible parties. For example, the lab safety official must have the authority to stop or shutdown work that is being conducted unsafely. Moreover, the lab safety official must have enough authority to prevent individuals from entering unsafe areas such as spill zones. In some environmental laboratories it is the lab safety officer who conducts the in-house safety

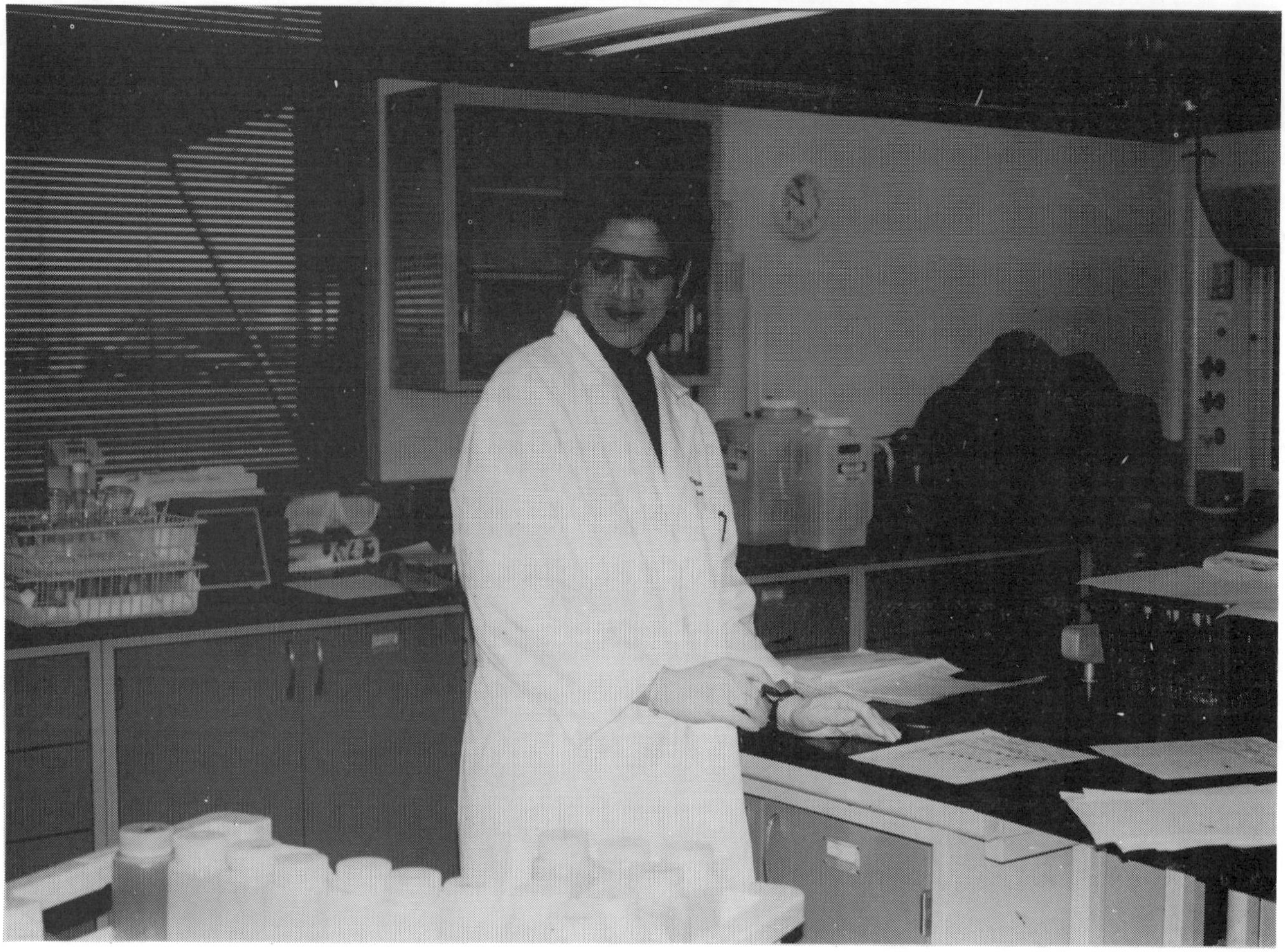

Figure 2.3. A lab specialist outfitted in proper personal protective equipment (PPE).

audits. The safety officer must have enough authority to conduct a proper audit without threat of harassment from those being audited. The auditor must also be able to put a stop to or to correct unsafe work practices as they occur. For example, the lab safety officer might discover during an audit (or any other time) that a lab worker is using an electrical extension cord that is frayed, broken, cut, or damaged in some way. The safety official should (must) have the authority to stop the use of this electrical hazard.

Although the type of authority described in the preceding is critical in allowing the laboratory safety officer to correctly perform his/her job responsibilities, it should be pointed out that this authority is *latent* authority. That is, this type of authority is reserved and should only be used with great care and discretion. The laboratory safety officer must be an enforcer of good order and discipline in regards to performing job tasks safely in the laboratory—he/she should not act as a Gestapo agent.

Experience has shown that one thing is certain in regards to the laboratory safety officer's responsibilities and authority: If the laboratory safety officer is not given adequate authority, he/she will not be able to meet his/her assigned responsibilities.

2.5 LABORATORY SAFETY RULES

Earlier it was pointed out that laboratory supervisors are responsible to ensure that safety rules are established for the lab and that laboratory workers know and follow the safety rules. While it is true that it is impossible to design a set of safety rules that will cover all possible hazards and occurrences, it is also true that worker safety awareness can aid in this area. As stated earlier rules are everywhere and most people do not like rules. In order to reasonably expect lab workers to perform their work in accordance with lab safety rules several conditions must be met:

(1) Laboratory safety rules must be posted.
(2) Laboratory safety rules must be written in simple form.

(3) Training on laboratory safety rules must be conducted.
(4) Enforcement of laboratory rules must be consistent.
(5) Do not allow familiarity with lab operations to lead to the overlooking or underestimation of hazards.
(6) Laboratory supervision must set the proper example.

The last condition listed above is important. Whenever a supervisor violates an organization's rules, then it is to be expected that workers observing these violations will perform in a similar manner. As a case in point, let's look at a situation that might exist in an environmental laboratory.

EXAMPLE

The laboratory supervisor is a nice guy; he smiles a lot—a Good Neighbor Sam-type. He feels that the 95 lab technicians that work for him are the best that can be found anywhere. This particular laboratory supervisor is also well-experienced, knowledgeable—well-educated. He has a problem, however. He understands the need for laboratory safety rules but feels that they are burdensome, designed for others to worry about, and not important enough to merit enforcement.

"Why worry about someone's safety and health," he tells himself (on a daily basis), "especially in the lab—my people are all college graduates, professional folks—they know how to protect their safety and health. Anyway, all any of my folks have to do is to follow my lead, then they won't mess up and get hurt. Don't you think?"

On any given workday in his laboratory, the lab supervisor does wear his lab coat. However, he wears the lab coat not because it is required by lab safety rules but because he does not want his expensive personal clothing to get dirty or destroyed by lab chemicals. His lab coat can be buttoned (as per lab rules, this is a requirement); however, he chooses not to button his coat because it's too much of a hassle.

Along with the rule about wearing a fully-buttoned lab coat while in the lab, this particular lab has several other safety rules. For example, there is a rule against

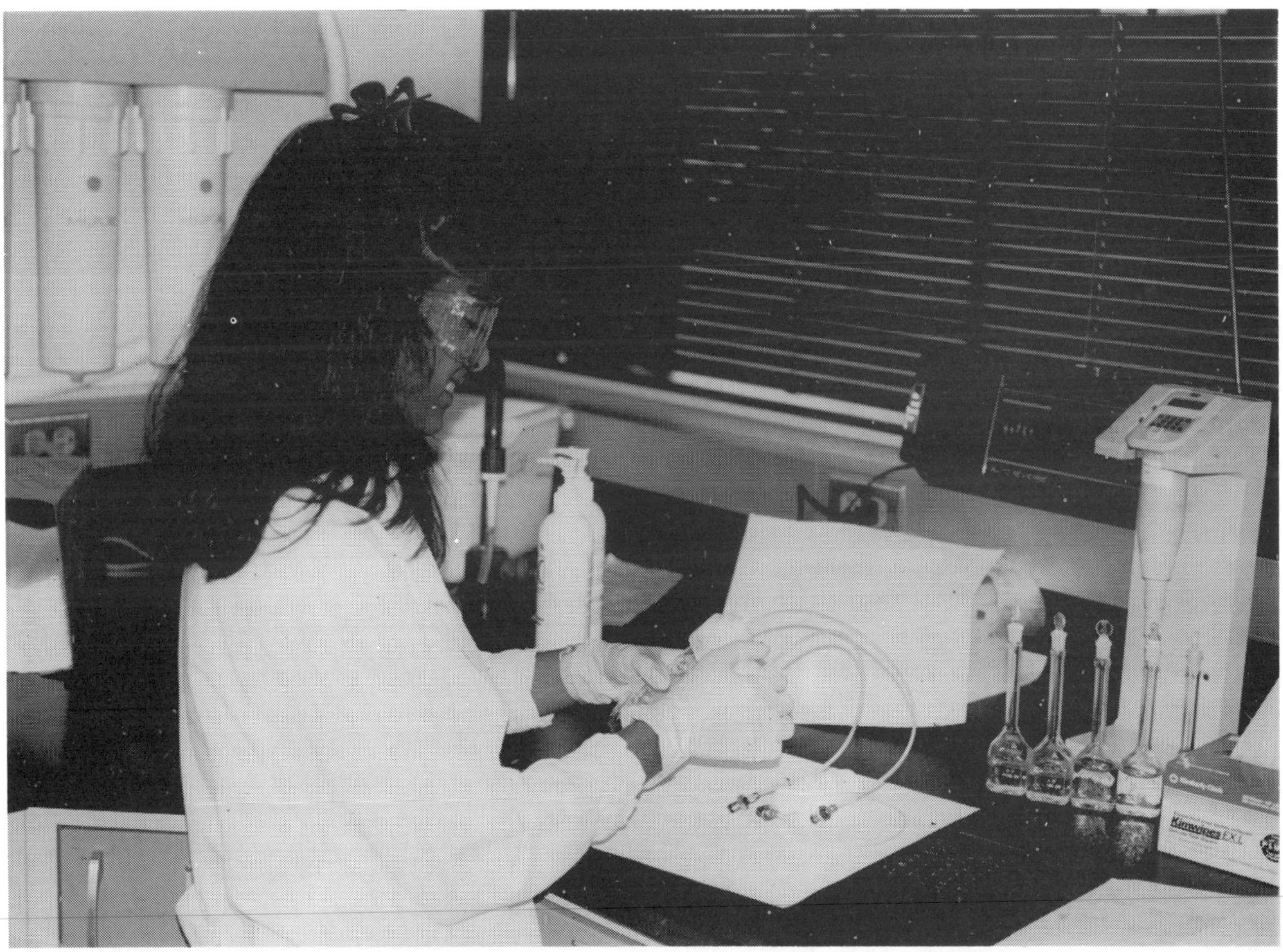

Figure 2.4. Another lab specialist outfitted in proper personal protective equipment (PPE).

smoking cigarettes anywhere inside the laboratory building. The lab supervisor is aware of this rule (he ought to be, he wrote the rule) but chooses to ignore it.

"I'm in charge around here—rules do not apply to me," he reminds himself.

"Besides, I'm the guy who made this place a smoke-free work place—a smoke-free workplace is good for my workers' morale. I believe in a smoke free workplace and I expect my workers to abide by the rules—but in my case, well, supervisors should have some privileges—don't you think?"

When the lab supervisor isn't walking around with his open lab coat and a cigarette dangling from his mouth, he is sometimes asked for guidance by his workers. For example, his workers might want to know what they should do with waste materials they produce when performing standard laboratory analysis using chemicals. These wastes usually consist of mercury, phenol, and other chemical mixtures. Various regulatory agencies have made it perfectly clear that chemical waste products must be properly disposed. The lab supervisor knows the rules, the requirements, but he also knows that properly disposing chemical wastes is burdensome and time-consuming. Further, he knows that proper disposal of hazardous chemical waste is expensive—to his way of thinking, too expensive. Therefore, whenever he is asked about disposing laboratory wastes, he gives the same standard answer: "You know what to do with it." Thus, the lab workers pour the waste down the sink drains. This out-of-sight-out-of-mind work practice has become standard operating procedure for this particular lab.

As stated earlier, the lab supervisor is a really nice guy; he smiles a lot. In his own mind he is a legend; liked by all his serfs. On a day to day basis he literally goes out of his way to keep his 95 lab technicians happy. Thus, whenever he spots a lab worker violating one of the lab's safety rules, such as finding a worker's lunch in the lab refrigerator, he ignores it. Moreover, whenever a lab worker places broken lab glass in lab garbage cans, he ignores it. Whenever lab workers joke and push each other around, he ignores them. Whenever he sees an unmarked chemical container, he ignores it. Whenever one of his lab workers violates any safety rule, he ignores it—he is able to justify their actions—in his own mind, that is. His thought process usually runs along the line of "Why should I hassle my workers about micro-minutiae—they get the job done—makes sense, don't you think?"

The most important ingredient in the mixture that gels to form a viable safety program is the mindset of the laboratory workers. Obviously, if the lab workers' mindset is formed through the example of the lab supervisor in the previous example, then the laboratory itself is nothing more than a disaster location waiting for the big event—the one that is bound to harm workers, shutdown the lab, and damage the environment. The previous example simply reinforces the old adage of ***leadership by example***. Further, the example this particular lab supervisor has set and demonstrated simply amplifies an even older fact. That is, an organization can develop all the safety rules it chooses to, but if these rules are not followed or abided by *all* personnel involved (including supervisors), then written safety rules are nothing more than empty words. We can only hope that the style of "***leadership***" demonstrated by the lab supervisor is rare—very rare. However, experience has shown that in the real world examples like this one are more common than I like to admit.

When it comes to formulating the actual set of safety rules to be used in the environmental laboratory, careful thought should be given to this process. There are several sets of "canned" safety rules that one might choose to use in his/her laboratory. Again, it is important to keep these rules limited to a small, concise, clearly written set rather than producing a voluminous document that no one will bother to read.

Safety rules intended for the environmental laboratory should, at the minimum, consist of a compilation of general safety rules, rules pertaining to personal hygiene, glassware safety, and housekeeping safety rules. In the following section an example of the type of safety rules that are commonly found in environmental laboratories are presented. These rules are limited in number for two reasons: (1) the smaller the number of safety rules the more likely they are to be known and observed by all lab workers; and (2) limiting the lab safety rules to a small number allows for use of a single sheet of paper or poster board. A single one-sided set of safety rules is easier to post and to read.

2.5.1 SAFETY RULES FOR THE ENVIRONMENTAL LABORATORY—AN EXAMPLE

Every laboratory worker must observe the following safety rules:

(1) All work areas and laboratory benches must be kept clean and clear of clutter.

(2) All laboratory workers must wear lab coats and use appropriate personal protective equipment (PPE) while in the laboratory.

(3) Avoid damaging any laboratory glassware by following careful handling and storage procedures. Discard all damaged glassware items into designated containers.

(4) Follow the safety rules and procedures that apply to each laboratory operation.

(5) Be alert to unsafe conditions or hazards. Report all hazards *immediately*.
(6) Avoid mouth suction to pipet chemicals or to start a siphon. A pipet bulb or aspirator should be used to provide vacuum.
(7) Hands are to be properly protected when inserting glass tubing into rubber stoppers or corks on when placing rubber tubing on glass hose connections. Glass tubing must be fire-polished or ground round and lubricated.
(8) All working surfaces and floors must be cleaned regularly.
(9) Hands must be properly protected when picking up any broken glass.
(10) Access to all emergency equipment, showers, eyewashes, and exits must never be blocked by anything.
(11) Use laboratory equipment only for its designed purpose.
(12) All chemical containers must be properly labeled with the identity of the contents and the hazards presented to users.

A final point on laboratory safety rules is made here before moving on to the next topic. Past experience has shown that when general housekeeping in the laboratory is poor, the lab's overall safety profile is also lacking. The point is that there is a definite, direct relationship between orderliness in the laboratory (or other workcenters) and its safety profile.

2.6 OSHA'S LABORATORY STANDARD

In the not too distant past it was common practice for academic, industrial, and other laboratories that used hazardous chemicals/materials to follow "homemade" or generic safety standards. These generic safety standards or procedures were usually developed by someone who was designated responsible for safety in the lab. From laboratory to laboratory the content of these safety procedures varied. When OSHA began to publish its various safety standards, some of these new regulations were implemented for use in laboratory operations. Just a few years ago, probably the most commonly used OSHA standard was its 29 CFR 1910.1200 Hazard Communication (HazCom) Standard. HazCom requires facilities where hazardous chemicals/materials are used, handled, or produced to implement a HazCom Program. Under the HazCom Program, the affected facility is required to have a written program, provide employee training, use an easily recognizable labeling system for all hazardous chemicals/materials, and to make available to all employees, 24-hours per day, Material Safety Data Sheets (MSDS) for each hazardous chemical/material used in the workplace. HazCom has been an extremely beneficial program that has worked to enhance the safety of workers in the laboratory (it was the foundational document that made up the laboratory safety program).

Eventually, OSHA came to the conclusion that laboratories using hazardous chemicals/materials were somewhat different (more complex) than other types of work centers that also used these materials. Further, laboratory workers are sometimes exposed to a wider variety of hazardous materials and work activities than the average industrial worker. Therefore, in order to establish consistent safety guidelines for use in laboratory environments ***only***, OSHA, in 1992, promulgated 29 CFR 1910.1450, "**Occupational Exposures to Hazardous Chemicals in the Laboratory**." This standard is now commonly known as the *Laboratory Standard*.

Note: Laboratory managers should not confuse the relationship between OSHA's Laboratory Standard and Hazard Communication Program. As pointed out earlier, OSHA requires employers who expose workers to hazardous chemicals in the workplace to abide by requirements covered under HazCom. The Laboratory Standard adopted specific parts of HazCom and incorporated these parts into its requirements. The point is that the Laboratory Standard still requires labeling, retention of MSDS received with chemicals, and worker training.

OSHA's Laboratory Standard is a *performance standard*. Performance standards indicate the end result to be achieved, but not the methodology by which it is to be accomplished. In plain English, what a performance standard does is give the employer a certain amount of "*license*" to achieve compliance with the requirements of the standard. The primary emphasis of a performance standard is on implementation of administrative controls designed to protect workers from over-exposure to hazardous chemicals/materials in the laboratory.

Two important requirements of OSHA's Laboratory Standard are presented as follows:

(1) Laboratories that use hazardous chemicals in their spaces where workers might be exposed to chemicals must follow the guidelines provided in the Laboratory Standard. Keep in mind that "workers" can include such personnel as those who work in the lab office, maintenance and custodial personnel, and others who regularly spend a significant amount of their time within the lab environment.
(2) The Lab Standard requires laboratories that use hazardous chemicals to develop and implement Chemical Hygiene Plans (CHP). The CHP must outline the specific procedures, safe work practices, and personal protective equipment that are required to be used in

the laboratory to control occupational exposures. It is important to point out that all personnel associated with laboratory operation must share chemical hygiene responsibilities. More will be said about the Chemical Hygiene Plan and its requirements later.

2.7 ACCIDENT/INCIDENT REPORTING PROCEDURE

No matter how well organized and effective the Environmental Laboratory's safety program is, worker injuries, chemicals spills, and exposure to biologicals will occur. In the Environmental Lab, where exposure to all sorts of chemical and biological hazards is a routine occurrence, even the most minor scratch, cut, and abrasion should immediately be reported to the lab supervisor. Some would say that it is burdensome for the lab supervisor and the lab safety officer to process paperwork that details minor on-the-job injuries. This might be the case in other industries and environments, but in the Environmental Laboratory, it is important to require lab workers to report all injuries, no matter how minor.

Along with the need to report all accidents/incidents that occur in the environmental laboratory, it is also important to investigate these occurrences. A word of caution: these investigations should not be designed to point the finger of blame at anyone. While it is true that when misbehavior or incompetence causes an accident or incident to occur in the environmental laboratory, these occurrences must be thoroughly investigated and remedial action administered swiftly, it is also true that investigations dealing with accidents/incidents that occur due to other causes should be fact-finding actions only. The goal of the investigation should not be to embarrass or chastise anyone. Instead, the goal must be to determine the causal factors and to correct them as soon as possible.

The laboratory accident/incident investigation should be carried out by an independent person. For example, if the environmental laboratory is part of a larger organization that includes a separate and independent safety division, then safety division personnel should conduct the investigation. On the other hand, if the laboratory is a single, complete entity, then it is prudent to use the in-house team approach. This in-house accident/incident investigative review team should be independent and capable of conducting an effective evaluation.

In regards to the accident/incident team's expertise (capabilities) in conducting the review or investigation, it is important to ensure that team members first receive training on how to conduct an effective accident/incident investigation. Experience has shown that when accident/investigation training becomes part of the organization's overall training plan, then several informed personnel are always available to act as members of such a team. More will be said about this important training element later in this text.

TABLE 2.2. Laboratory Accident/Incident Investigation Form Required Information Elements.

1. Location of accident/incident.
2. Date of accident/incident.
3. Time of accident/incident.
4. Date accident/incident reported.
5. Time accident/incident reported.
6. Type of accident/incident (e.g., fire, spill, explosion, employee exposure, other).
7. Name of investigator/investigation team members.
8. Description of accident/incident. It is also important to include which operation or task was being performed at time of accident/incident.
9. List all potential causal factors.
10. Describe any exposures, illnesses, or injuries that may have occurred.
11. Describe which hazard control measures (e.g., fume hoods, ventilation, etc) were operating at the time of the accident/incident and the PPE worn.
12. Describe any safety Standard Operating Procedures (SOPs) or organizational policies that may apply to the accident/incident, including first aid and emergency response. Indicate if the proper procedures were being followed.
13. Check the level of training of those involved in the accident/incident. When was the last time training was conducted?
14. Recommendation(s) on how this accident/incident could have been avoided.
15. Management review

When the safety official or assigned investigation team conducts the accident/incident investigation it is important to ensure that the findings of the investigation are prepared in writing. The environmental laboratory should develop a standardized Laboratory Accident/Incident Investigation Form. This standardized form should contain information elements similar to those shown in Table 2.2.

It is important to take a moment here and discuss the last element listed in Table 2.2—**Management review**. Whether one is conducting a safety audit or an accident/incident investigation, either of these two important functions is a useless, wasted effort, unless the unsatisfactory items/casual factors found are corrected. The key word here is FOLLOWUP, FOLLOWUP, FOLLOWUP. Without followup, usually nothing gets accomplished. On the other hand, when management is required to review the results of safety audits and accident/incident investigations, this followup procedure usually snowballs into to additional followup. That is, the findings are acted upon and the discrepancies, unsafe work practices, incompetence, misbehavior, or other causal factors are corrected. When you get right down to it, isn't this what safety audits and/

or accident/incident investigations are all about? More specifically, remember, the target is the **hazard** that causes the unwanted harmful event—it is the *hazard* that must be eliminated.

Note: In the preceding section the reader has undoubtedly noticed the writer's use of the terms accident/incident in describing various actions and forms. The accident/incident label is deliberate. In the real world of safety practice it has, unfortunately, been the tendency of many practitioners to use the terms accident or incident interchangeably. As pointed out this practice is unfortunate. The two terms, of course, have different meanings. First, an *accident* is, according to Webster's definition, "an event or condition occurring by chance or arising from an unknown or remote cause." When the safety auditor or investigator lists an injury, mishap, or action that causes damage as an accident, this may or may not be the case. Consider the following case, for example.

A lab technician, who washes a lab instrument in 60% acid solution and wears no hand protection while doing so, will be said by some to have had an accident when burned in the inevitable skin- corrosive substance event this practice will cause. Others will say that the event could not be termed an accident, since it was predictable and therefore not unexpected.

An *incident* is not an accident. More precisely (correctly) an incident is an occurrence or event that is related or dependent on another thing. It is this "other thing" (i.e., the hazard or unsafe work practice) that the safety auditor or investigator is attempting to identify.

2.8 SAFETY AUDITS IN THE ENVIRONMENTAL LABORATORY

Safety audits or inspections can be a valuable tool in detecting laboratory hazards that may lead to worker injury, chemical spill, or fire in laboratory. The obvious purpose of safety audits is to identify and correct laboratory hazards.

A valuable tool when conducting the laboratory safety audit is a Safety Inspection Checklist. A safety checklist used for auditing the Environmental Laboratory should a be a living document; it should continue to grow as time passes.

Under OSHA's Laboratory Standard, environmental laboratories are required to institute a Chemical Hygiene Plan (to be discussed in detail later). One of the elements that should be included in the Chemical Hygiene Plan is a set of protocols for auditing and inspecting the laboratory.

Hazard evaluation and identification (the purpose of conducting workplace audits or inspections) will be discussed in greater detail in Chapter 3. However, a final point about auditing and inspecting should be made here before moving on. When conducting the safety audit/inspection in the environmental laboratory (or in any workplace) the inspector should keep in mind that his/her purpose is not only to identify hazards but also to ensure that these hazards are mitigated. Additionally, the laboratory safety auditor/inspector should try to view the laboratory environment as an outside inspector would. If the in-house safety auditor/inspector is able to take this view, then whenever an outside regulator does show up to inspect your laboratory, the laboratory will be ready for inspection; it should pass inspection with flying colors. When you get right down to it, isn't it always better to be commended than to be lambasted, especially when it is issued by a federal, state, or local regulatory official?

Hazard Evaluation and Identification

> Hazard evaluation and identification are essential if the laboratory safety program is to function properly. If there is no effort to ensure that appropriate procedures and policies have been carried out conscientiously, the lab's safety program may exist only on paper. (National Academy Press, 1989, p. 127)

3.1 INTRODUCTION

In Chapter 1 it was pointed out that OSHA, in its **Guidelines on Workplace Safety and Health Program Management (54 CFR 3908)**, (Laboratory Standard), outlined four elements that makeup a standard health and safety management program. The second of these elements, ***work site analysis***, is described in this chapter.

3.2 HAZARD EVALUATION AND IDENTIFICATION

Before an environmental laboratory is built, considerable effort is expended in the planning and design stage. This planning and design stage ensures that various construction codes and regulations are identified—to be followed in the construction process. More importantly, to ensure the safety and good health of laboratory personnel who are to work in the laboratory, engineering controls are specified and then installed. These engineering controls usually include the following:

(1) Machine guards on all rotating equipment
(2) Installed ventilation and fume hoods
(3) Handwashing facilities/materials
(4) Safety showers located throughout the facility as required by local, state, and federal codes and regulations
(5) Restricted access and isolation
(6) Barrier systems where required

With all the effort expended and expertise utilized in designing and then building a first-class environmental laboratory, one might logically assume that the facility would be safe for use by laboratory personnel. This reasoning (assumption) is correct to a point. However, another vital aspect must be considered. For example, if one purchases a brand new automobile, it can logically be assumed that it was designed and built with personal safety in mind; that is, the automobile is designed and built to be as safe as it can be. However, this automobile with all of its engineering controls designed to protect the good health and safety of the driver is of little good if the driver ignores good driving practices. Likewise, designers, engineers, and builders can design, engineer, and build an environmental laboratory that is safe, but if those who use this facility do not follow safe work practices, then built-in safety devices and engineering controls are useless.

At this point, the reader might be asking him/herself: What does the lack of safe work practices in the laboratory have to do with hazard evaluation and identification?

Simply stated, *everything*!

The point is that safe work practices performed on a consistent basis by laboratory technicians (and any other worker) is the key to safety in general—anywhere, anytime. Performing one's tasks while utilizing safe work practices does not eliminate the need for hazard evaluation and identification, but makes the task easier and more beneficial.

3.3 AUDITS/INSPECTIONS

As stated previously, in order to maintain a safe environment in the environmental laboratory or in any other workplace, workers must be trained to perform their tasks safely; that is, they must practice safe work practices.

Unfortunately, performing work utilizing safe work practices is not automatic. Safe work practices must be learned. Learning, although critically important, is never the total answer. For example, one can learn how to repair a gun safely—following safe work practices. The problem is that not everyone will do work, like repairing a gun, following safe work practices.

In order to ensure that safe work practices are being practiced on a daily routine basis it is necessary to institute a quality assurance program in the workplace. This quality assurance program can be assured via laboratory safety audits/inspections.

It should be pointed out that an effective safety audit/inspection *must* include four key elements:

(1) The safety audit or inspection procedure must be more than just a weekly, monthly, or quarterly formal event; it must be conducted informally on a daily basis. Not only should the laboratory supervisor, safety officer, or chemical hygiene officer make daily walk-arounds to check on proper performance of safe work practices but each lab technician and worker should also make daily rounds.

(2) It does little good to attempt enactment of quality assurance in any workplace unless those doing the safety audit/inspection are knowledgeable about work practices and how to identify hazards associated with such practices.

(3) The workplace audit of inspection is difficult to conduct without using some type of checklist (see Table 3.1). Since each laboratory is different, the checklist should be site-specific. It should be pointed out, however, that no site-specific, generic checklist can include all safety and/or health hazards. Therefore, the workplace audit/inspection checklist serves as a reference tool only.

(4) The results or findings of the safety audit/inspection should be documented. This documented record should be maintained on file for future reference. Further, subsequent audits/inspections should include the use of previous audit/inspection records for followup. Followup ensures that all previous discrepancies are cleared. Remember, it is the purpose of safety audits/inspections to identify all hazards and to ensure that they are *cleared.*

3.3.1 AUDIT/INSPECTION CHECKLIST

This section presents an example of an inspection checklist (see Table 3.1) that has been used by a large environmental laboratory (more than 90 personnel). It is important to point out that this particular checklist is only a sample. Each environmental laboratory should develop a site-specific checklist that addresses the types of hazards that might be present in the facility to be audited/inspected. The checklist must also reflect or consider site policies, procedures and practices. Additionally, whatever type of safety audit/inspection checklist that is developed, it is important not to keep this document secret. To the contrary, this audit/inspection checklist should become a standard workplace training aid. The point is that every lab worker should be intimately familiar with every laboratory safe work practice and potential hazard. When you get right down to it, how else are lab personnel to be expected to ensure their own (and others) health and safety while performing work in the environmental laboratory?

TABLE 3.1. Sample Laboratory Safety Inspection Checklist.

item	Yes/No	Comments
1. Food and beverages are not stored in laboratory refrigerators or laboratory area.		
2. Hazardous chemicals are used only in laboratory hoods or other designated area.		
3. Required PPE is utilized.		
4. Laboratory surfaces are cleaned (disinfected) after work is completed and after any spill.		
5. No smoking, food, and beverage signs are posted and rules are observed.		
6. Lab coats are available and properly used.		
7. Respirators are available and properly used.		
8. All containers containing hazardous chemicals/materials are properly labeled.		
9. All laboratory storage areas, lockers, and bins are properly labeled.		
10. MSDS are available for all chemicals used or stored in the lab.		
11. Laboratory bench tops, areas under sinks, and floors are orderly.		
12. Aisles and exits are free from obstructions.		
13. Electrical equipment and cords are in good condition.		
14. Tools and equipment are in good repair and electrically grounded.		
15. Tops of shelves or cabinets are free of stored items.		
16. Glassware is free of cracks, sharp edges, and other defects.		
17. Areas using etiologic agents (biohazards) are properly posted and entry restrictions are being followed.		
18. Incompatible materials are properly segregated.		
19. Hazardous materials used/stored in the laboratory are limited to small quantities.		
20. Safety carriers are available for glassware.		
21. Surplus chemicals are being disposed of in accordance with regulatory requirements.		
22. Compressed gas cylinders are properly secured/chained.		
23. Full and empty gas cylinders are stored separately.		
24. Propane/acetylene gas is separated from oxygen by at least 20 feet.		
25. No chemicals are poured down the sink or drain.		
26. Hazardous wastes are not accumulated for longer than regulations specify.		
27. Flammable liquids are stored away from sources of ignition.		
28. Corrosives and flammables are stored below eye level.		
29. Solvent waste cans are properly labeled.		
30. Empty gas cylinders are labeled as such.		
31. Syringes and other sharps are disposed in proper container.		
32. Bulk quantities of flammable liquids are stored in approved storage cabinets in required amounts.		
33. Volatile liquids are kept away from heat, sun and other sources of heat.		
34. Flammable liquid storage lockers are properly labeled.		
35. Gas cylinders are properly labeled as to their contents.		
36. Waste material is not allowed to accumulate on the floors, in corners, or under shelves and tables.		
37. Compressed gas cylinder caps are in place when cylinders are being moved or are not in use.		
38. Compressed gas cylinders are stored away from excessive heat.		
39. Solid and liquid wastes are separated.		
40. Hazardous wastes are not disposed of in the sewer system.		
41. Safety cans used to handle small quantities of flammable liquids are properly labeled.		
42. Safety showers & eyewashes are located in, or close proximity to, each area where hazardous chemicals are used or stored.		
43. Warning signs or labels are posted wherever required.		
44. General ventilation meets all regulatory requirements.		
45. Hoods are in good working condition.		

TABLE 3.1. (continued).

item	Yes/No	Comments
46. Hoods are not cluttered with equipment.		
47. Exits are clearly marked and visible.		
48. Each laboratory is equipped with a sink for handwashing.		
49. All fire doors are self-closing and are kept closed.		
50. Fire alarms are provided where necessary.		
51. Each fume hood is marked with operating heights, average face velocity, and any restrictions for use.		
52. All solvent cabinets are provided with exhaust ventilation.		
53. Safety showers and eyewashes are clearly labeled, and these areas are clear from obstructions.		
54. Spill containment/clean up material is readily available and clearly labeled.		
55. The eyewashes and safety showers are in good working condition with inspection tag posted.		
56. Emergency evacuation routes are clearly posted.		
57. Neutralizers are located in areas where acids are poured or handled.		
58. First aid supplies are readily available and clearly labeled.		
59. Biological wastes are disposed of in containers labeled with the appropriate biohazard symbol.		
60. Proper procedures for storage and disposal of biohazards are being followed.		

PART II

LABORATORY ADMINISTRATION REQUIREMENTS

CHAPTER 4

Documentation Requirements

4.1 INTRODUCTION

DOCUMENTATION requirements, pertaining to workers' safety and health, for any industrial work center subject to the Occupational Safety and Health Act of 1970 can be quite extensive. This becomes clear when one reviews the various OSHA standards. Under OSHA, three documentation requirements are consistent with each standard: record keeping, written safety and health programs, and worker training. This section discusses the record keeping and written program requirements for the typical environmental laboratory. Worker training requirements will be addressed later in this text.

4.2 DOCUMENTATION REQUIREMENTS

Environmental laboratories are required to maintain records for worker safety and health training, medical surveillance, accident/incident reports, and audit/investigation reports. In addition to these standard requirements, the environmental laboratory is also required to fashion and/or maintain written Lab Health and Safety Plans, Standard Operating Procedures, a Respiratory Protection Program, Health and Safety Information Centers, Exposure and Monitoring Records, and Hazardous Waste Reporting Procedures and Records.

The requirement for a Laboratory Health and Safety Plan and Standard Operating Procedures (SOPs) will be discussed in this chapter.

4.3 HEALTH AND SAFETY GUIDELINES FOR THE ENVIRONMENTAL LABORATORY

4.3.1 HEALTH AND SAFETY PLAN

At the present time, most environmental laboratory workers associate their lab's Health and Safety Plan with the lab's Chemical Hygiene Plan (CHP), which is an OSHA requirement. This association evolved around the approved practice of substituting the CHP for the lab Health and Safety Plan. This substitution process is common practice because under OSHA requirements a CHP is required even if the lab already has a Health and Safety Plan; thus, it is more practical to use the Chemical Hygiene Plan alone, instead of both plans. *A word of caution:* OSHA only allows a lab to substitute the CHP for the Health and Safety Plan on condition that certain requirements are addressed.

In addition to the Standard Operating Procedures (SOPs) (to be discussed later), the Chemical Hygiene Plan should address:

- general housekeeping
- eating and smoking areas
- emergency procedures
- health and safety organization and policies
- warning signs and labels
- personal protective equipment (PPE)
- engineering controls
- fire protection and prevention
- laboratory safety audit/inspection
- medical surveillance
- employee training
- waste disposal
- chemical storage
- respiratory protection program
- record keeping
- personal and environmental monitoring
- location of wash stations and emergency showers

4.3.2 STANDARD OPERATING PROCEDURES

In developing standard operating procedures (SOPs) for the environmental laboratory, some thought should go into (1) choosing the format to be used and (2) in selecting the actual elements that will make up the SOP. In regards to the SOP's format, it has become standard procedure to use formal company/organizational stationary. This makes sense when you consider that SOPs are official documents. On this official laboratory stationary (letterhead) should be included (1) the company name (laboratory name); (2) the SOP topic or subject; (3) the issue or revision date; (4) both the issuing individual and the official approving the SOP should sign the form; and (5) each section number and page number should be used.

The first step in putting together the SOP is to list each step-by-step procedure. After these step-by-step procedures have been listed, elements such as data, graphs, flow charts, equipment to be used during the procedure, and list of references used to develop the SOP should be included. Keep in mind that the actual elements making up the environmental lab's SOP will vary from facility to facility.

In regards to the references available for use in developing the lab's SOP, it should be pointed out that, depending on the function of the lab (i.e., clinical, nonchemical, research, or environmental), there are several available. Probably the two most commonly used in this process are USEPA's 40 CFR, Part 722, Subpart E (lists ***good laboratory practice*** [GLP] related to SOPs) and 40 CFR, Part 160, Subpart E (normally used for pesticides and provides excellent guidelines for GLP in the laboratory).

The environmental laboratory should have written SOPs for the following activities:

- hazardous waste disposal
- emergency response procedures for chemical spills, fires, and medical emergencies
- spill cleanup procedures
- hazardous chemical handling
- ventilation system maintenance
- respiratory protection and fit test
- visitor access to lab areas
- employee training
- personal protective equipment (PPE)
- housekeeping practices
- transport, storage, receipt, and shipping of hazardous substances
- medical surveillance
- eye protection

4.4 ESTABLISHING THE ENVIRONMENTAL LABORATORY DOCUMENTATION PROGRAM

When in the process of establishing a laboratory safety & health program, it is important for the lab's safety officer/ chemical hygiene officer to review pertinent safety & health record keeping requirements. This is especially the case in this age of spontaneous, at the drop of a hat litigation practices. The point is it is imperative that employers maintain accurate and complete records of their health and safety programs.

Along with the need to protect against potential litigation there are several regulatory requirements that must be met and abided by; failure to comply with such regulations can lead to steep fines, incarceration, and shutdown of the laboratory. It is also important to keep in mind that environmental laboratories must also comply with all pertinent local, state, and federal regulations, many of which require employers to maintain very specific records.

To facilitate the lab's documentation and record keeping program, standardized documentation should be established and maintained. When properly established, a documentation program will create a paper trail that will provide regulators with an accurate "picture" (or profile) of the facility's compliance status.

After the lab's record keeping procedure and written programs have been developed, it is important to ensure that all lab workers are familiar with these documents. Managers sometimes forget to include workers in the regulatory compliance effort. This mistake is common and sometimes costly. When an entire organization is properly informed—properly tuned-in and communicating—it is better able to implement, cope with, and abide by regulatory requirements.

It is also important to maintain current records, programs, and other documents. Regulatory requirements change frequently—management must stay on top of these changes when they occur and take necessary steps to update and maintain documentation in the laboratory. Because of occasional changes in operating procedures and equipment, it is also important to update Standard Operating Procedures.

Experience has shown that it is wise to develop SOPs for as many laboratory functions as possible; they should not merely describe a particular work activity. Instead, SOPs should provide step-by-step procedures on *how* to perform a particular function or procedure.

CHAPTER 5

Training Requirements

5.1 INTRODUCTION

TRAINING is at the heart of safety. Without a viable training program it is almost impossible to establish and maintain a safe and healthy work environment. Moreover, without training it is almost impossible for workers to perform their functions in a safe manner. Worker training goes beyond safety concerns, managers must also worry about workers being able to perform their work correctly, efficiently, and in a timely fashion. One thing is certain, if workers are not properly trained to perform their assigned tasks, then it is almost certain that their work will be done incorrectly, inefficiently, and in an untimely fashion.

Most of the training requirements that will be discussed throughout this text are mandated by OSHA, EPA, and DOT; environmental laboratories must comply with these mandates. It is interesting that more than 100 OSHA, DOT, EPA safety and health regulations contain training requirements. It is even more interesting to note that OSHA, for example, although it requires extensive safety and health training, does not always specify what is required of the employer who is required to provide the training.

OSHA allows employers a certain amount of license in accomplishing the employee safety and health training requirements. What is important for those who write the training programs to remember is the key point of training: Workers cannot be expected to perform their assigned tasks safely unless they are aware of the hazards or the potential hazards involved with each job assignment.

Experience has shown that if the lab's (or any other entity's) training program is not well-planned, employees will not be properly trained; lack of planning can also negatively influence their performance. This, of course, must be avoided.

A well-planned training program consists of three major components: (1) program needs and objectives are identified; (2) training outlines are developed and training aids gathered, assembled or procured; and (3) the training method employed is properly designed to fit the site-specific situation.

Without clearly defined training needs and objectives, it is difficult to design and then put together a viable training program. For the environmental laboratory, **training needs** should be accomplished when: (1) new employees are hired; (2) workers are transferred from one lab activity to another; (3) new processes or equipment are introduced; (4) SOPs are revised or when new ones are put on line; or (5) when safety performance shows a definite need for improvement—for refresher training.

Without clearly defined **objectives** it would be difficult to determine the exact scope of the training. Without determining the exact scope of the training it becomes impossible to select or prepare the training materials that will be needed.

Experience has demonstrated that even a well-designed training program is useless unless it is supported by *all* layers of management. Moreover, laboratory training should not be designed for the individual but instead for the group—including management personnel. The point is site-specific safety and health training should be for all those who work within the work center. Additionally, experience has shown that when all layers of management support and are included in an organizational training program, the proper climate is set to create the type of organizational culture in which training will flourish and, in turn, will enable performance to excel.

After program needs and training program objectives are defined, the next step is to develop an outline of the topics that are to be covered and a schedule of when training is to be presented. Today, many facilities utilize commercially available "canned" training. This canned

Figure 5.1. Employee training is an important part (the heart of) any safety program.

training is available in a wide variety of topics presented in audiovisual format. When this canned training is combined with site-specific in-house information, it can enhance the overall training effort.

The training methods employed in a particular facility will depend on several factors. For example, factors such as training content, size of the group to be trained, and the background (education level) of the trainee must be considered when determining the method of training to be employed. The training type to be employed must also be considered. The type of training employed usually depends on whether it is designed for individual or for group instruction (see Figure 5.1).

Whether the type of training employed is for individual or group presentation, one point is certain: Formal training should not be given without proper preparation. Specifically, proper preparation means that the instructor has a format or lesson plan that will deliver the intended information in a clear, orderly, and logical manner.

To demonstrate a type of training format that presents valuable and required information to workers in a clear, orderly, and logical manner, the following section is provided.

5.2 TRAINER'S LESSON GUIDE: AN EXAMPLE

The following lesson guide is used by Hampton Roads Sanitation District (HRSD), Virginia Beach, Virginia, to train wastewater workers on Laboratory Safety. HRSD uses this lesson guide initially for new employees and as a refresher course presented to all HRSD plant and environmental laboratory personnel once each year.

HAMPTON ROADS SANITATION DISTRICT
LABORATORY SAFETY
TRAINER'S LESSON GUIDE
A GUIDE TO SAFE WORK PRACTICE

I. INTRODUCTION

The Occupational Safety and Health Administration's 29 CFR 1910.1200 (Hazard Communication Standard) requires employers to provide training on the hazards of chemicals that employees are working with or may be exposed to. The District has various small laboratories and field projects that require employees who ordinarily are not laboratory workers to perform laboratory analysis. Moreover, OSHA's 29 CFR 1910.1450 (Laboratory Standard) requires *all* laboratory personnel to be trained in basic safety in regards to working in any laboratory. In order to comply with the OSHA Standard and to ensure that the District's workforce is knowledgeable about safety precautions to be followed whenever working with chemicals in a laboratory setting, this lesson guide is provided for the Work Center Supervisor's use in ensuring that standardized Laboratory Safety training is conducted.

II. DESCRIPTION

Laboratory means a facility where the "laboratory use of hazardous chemicals" occurs. It is a workplace where relatively small quantities of hazardous chemicals are used on a non-production basis. Depending on the chemicals or substances handled, working in a laboratory may expose you to the hazards of corrosive and poisonous chemicals and infectious microorganisms. Under these circumstances, you must be constantly alert of your actions and aware of techniques in safe handling of potentially dangerous materials. Whenever additional information on any chemical used in District laboratories is needed, refer to the Material Safety Data Sheet (MSDS) for that chemical.

III. GENERAL SAFETY PRECAUTIONS

1. Store chemicals in a safe place where they are not hazardous to personnel or property.
2. Label all chemical containers indicating the chemical name and date of preparation and/or container opening.
3. Check the labels on chemical containers before using to ensure that proper chemicals are selected for use.
4. Properly dispose of unlabeled or out of date chemicals.
5. Dispose of chemicals safely. Never throw away chemicals into common trash cans. Contact the Chief of Technical Services Division or Chief of Industrial Waste Division when the disposal of chemicals is required.
6. Read and learn the directions for each chemical's use and safety. This information is found on the chemical's Material Safety Data Sheet.
7. Follow directions carefully. *Never* mix chemicals randomly or indiscriminately.
8. Handle chemicals carefully when pouring or measuring to prevent spillage. The use of funnels helps to greatly reduce spillage.
9. Immediately clean up/contain spills according to the directions on the chemical's Material Safety Data Sheet.
10. Avoid personal contact with chemicals by using proper personal protective equipment (PPE).
11. Wear the correct glove for the chemical being handled. Refer to the chemical's Material Safety Data Sheet if there is any doubt about proper protection. Never touch chemicals with bare hands.
12. Ensure protective gloves are free of cracks and holes, and that they fit properly before handling any chemicals.
13. Do not place fingers into mouth, nose, ears, or eyes while handling chemicals. Further, do not eat, drink, smoke, nor apply make-up while handling chemicals.
14. Wash hands with a disinfectant soap after handling chemicals.
15. Wash off chemical spills on skin immediately with running water.
16. Avoid breathing chemical mists, dust, or vapors.
17. Provide positive ventilation to laboratory work areas.
18. Properly dispose of all broken, chipped, or cracked glassware.
19. Do not use laboratory glassware as coffee cups or food containers.
20. Use a suction bulb to pipette chemicals or wastewater. *Never* use the mouth to suck up a fluid or chemical in a pipette.
21. Use safety goggles and/or face shield when transferring, measuring, or whenever there is a potential for chemicals to be splashed in the eyes.
22. Use tongs or thermal gloves when handling hot equipment.
23. Use only properly grounded electrical equipment.
24. Always *add acid to water* and *not* water to acid.

25. Use carbon dioxide or dry chemical type fire extinguishers to control laboratory fires.
26. Ensure the laboratory is equipped with an eyewash and that the eyewash is functional.
27. Ensure that laboratory broken glassware is properly disposed.

IV. SAFE PROCEDURES

A. Infectious Materials

Sewage and sewage biosolids contain millions of bacteria, some of which are infectious and can cause disease such as tetanus, dysentery, poliomyelitis, or hepatitis. Always wash your hands with soap and water, particularly before handling food. Use rubber gloves when handling sewage samples. Cuts or abrasions in the skin can easily become infected.

B. Corrosive Chemicals

Sulfuric, hydrochloric, nitric, glacial acetic, perchloric acids, and chromic acid cleaning solutions quickly destroy human tissue, clothing, and wood. They are also extremely corrosive to metals and concrete. Use glassware or polyethylene containers. Always add acid to water, not vice versa. Avoid contact with metals. Pour and pipette acid carefully. Never pipette by mouth; use a pipette bulb. If spills occur, follow these safety procedures:

- Dilute the affected area with large portions of water. Clean up the diluted material.
- For spills on bench tops, wear gloves and dilute and squeegee into the sink.
- On a person, immediately wash off with large quantities of water.
- If spills of concentrated acid splash on your face, flood with large amounts of cold water. Notify the Supervisor.
- If acid gets into your eyes, use the emergency eye wash station.

Bases such as sodium hydroxide, potassium hydroxide, and ammonium hydroxide quickly digest/corrode skin, clothing, and leather. Always handle them with extreme care. Use glassware or polyethylene containers. Ammonium hydroxide is extremely irritating to the eyes and respiratory system. When pouring ammonium hydroxide, use a fume hood and operate with a ventilation fan. In case of an accident involving bases, wash the affected area with large quantities of water until the slippery feeling is eliminated. When handling chlorine gas solution, also use a fume hood and avoid vapor inhalation. Secure the hood cover to prevent vapor escape. Ferric salts and ferric chloride are very corrosive to metals. Avoid bodily contact with them and wash off any residue immediately.

C. Toxic Materials

Apply the following safety precautions when working with toxic materials:

- Read all toxicity warnings and antidote information on labels of laboratory chemicals. Learn about the materials that you are handling by reading their Material Safety Data Sheets.
- Avoid ingesting or inhaling any chemicals, especially cyanide, chromium, cadmium, and other heavy metal compounds.
- Use a vented hood when handling ammonium hydroxide, nitric acid, bromine, ammonia, aniline dyes, formaldehyde, chloroform, and carbon disulfide. Moreover, carbon tetrachloride is absorbed into the skin on contact and is cumulative.

D. Explosives or Flammable Materials

Gases such as acetylene and hydrogen, and liquids such as carbon disulfide, benzene, ethyl-ether, petroleum ether, and acetone all require careful handling. Always store them in accordance with proper fire prevention procedures, as listed on the Material Safety Data Sheet. Follow these procedures when handling explosive or flammable materials:

- Use a vented hood.
- Never use explosive or flammable materials near open flame or expose to heating elements.
- Do not distill to dryness, or explosive mixtures may result.

E. Glassware

Broken glassware leads to employees cutting themselves or unnecessary chemical exposure. Follow these simple safety procedures when handling glassware in the laboratory.

- Dispose of all broken or cracked glassware in the designated disposal container immediately. Chipped glassware may still be used if it can be fire polished to eliminate the sharp edges.
- Use care and caution when making rubber-to-glass connections. Support lengths of glass tubing while they are being inserted into rubber. Fire polish ends of glass tubing until smooth. Use a lubricant such as water—*never use grease or oil.* Use gloves when making such connections and hold tubing as close as possible to the end being inserted to prevent bending and breaking. Never try to force rubber tubing or stoppers from glassware; cut rubber as necessary to remove.

V. Laboratory Hoods

A laboratory type hood is a device that is enclosed on five sides with a movable sash or fixed partial sash enclosed on the remaining side. The hood is constructed and maintained to draw air from the laboratory or room; and allows chemical manipulations to be conducted in the enclosure without insertion of any portion of the employee's body other than hands and arms. The following safety procedures are to be followed when using a hood.

- Any operation that might result in the release of toxic chemical vapor or dust must be performed under a properly functioning hood. Use a hood or other forms of local ventilation when working with any appreciably volatile substances.
- Confirm adequate hood performance before each use.
- Keep the hood "on" at all times when it is not in active use if toxic substances are stored in it.
- Keep materials stored in hoods to a minimum and do not allow them to block air flow.
- When the hood is not in use keep the sash in the down position.

Confirm adequate hood performance before each use by referring to the air flow monitor on the side of the hood. If the hood does not have a built-in airflow monitor, use a hood air flow measuring device such as a velometer. Confirm hood performance with the sash in the down position. Hood face velocity must not be excessively turbulent. Typically, a hood face velocity of 60–100 linear feet per minute is adequate. Also check to see if the hood inspection sticker is current (inspected within the last three months). Do not use a malfunctioning hood and report any deficiency immediately to a supervisor.

A. Hood Inspection

On a quarterly basis the Safety Division inspects all laboratory hoods used throughout the District. Further, hood face velocities are measured using a calibrated velometer. Hoods must have a face velocity of 60–100 linear feet per minute to pass inspection. Hoods that pass inspection receive the following dated inspection sticker:

HOOD
INSPECTION

THIS VENTILATION
HOOD WAS INSPECTED
AND APPROVED FOR
USE ON ________
DATE

VI. ADDITIONAL TRAINING MATERIALS

HRSD's Safety Division has various video films covering Chemical and Laboratory safety.

(*Note:* The film entitled *Lab Safety: Handling Hazardous Chemicals* should be shown when presenting this training lesson. This film can be checked out of the Safety Division's Safety Film Library. Further, if more detailed training is needed on specific chemical hazards, contact the Safety Division. The Safety Division's "Safety Film Library" has a wide variety of video tapes covering several types of chemicals and their associated hazards.)

5.3 SAFETY AND HEALTH TRAINING PROGRAM

Safety and health training should begin for new employees right after they are hired. This is an important point because such regulatory standards as OSHA's 29 CFR 1910.1200 Hazard Communication Standard specifically state that no employee is to work in any workplace containing hazardous materials until the employee has been thoroughly briefed on the dangers of hazardous materials. Therefore, since the standard environmental laboratory usually contains all types of hazardous materials, it is important to train a new employee before he/she is put to work in the lab.

5.3.1 NEW EMPLOYEE SAFETY ORIENTATION

As stated previously, no new employee should be put to work in the environmental laboratory until having been properly trained on the safety and health aspects of his/her job classification and new environment.

In order to ensure new lab employees are properly trained it is necessary for the laboratory's designated training official (usually the chemical hygiene officer) to devise some type of training curriculum, plan, or guide that will list the safety and health items each employee must be trained on. Moreover, new laboratory personnel who receive, store, weigh, dilute, package, or transport hazardous chemicals/materials must be thoroughly familiar with the lab's Right-to-Know station(s) and the Material Safety Data Sheets that must be available at each station.

New employee safety training for laboratory personnel should include requirements outlined in the following regulatory standards:

- Respiratory Protection, 29 CFR 1910.134
- Lockout/Tagout Procedures, 29 CFR 1910.147
- Hazard Communication, 29 CFR 1910.1200
- Hearing Protection, 29 CFR 1910.95
- Bloodborne Pathogen Standard, 29 CFR 1910.1030
- Portable Fire Extinguishers, 29 CFR 1910.157
- Personal Protective Equipment, 29 CFR 1910.132-138
- Emergency Plans and Fire Prevention Plans 29 CFR 1910.38
- Occupational Exposure to Hazardous Chemicals in Laboratories, 29 CFR 1910.1450

The training should also include an introduction to the site-specific Chemical Hygiene Plan. In addition, each new employee must be indoctrinated on the lab's safe work practices for each job function. These safe work practices are sometimes automatically included in well-conceived and well-written standard operating procedures. Obviously, a new laboratory employee should not begin work without first knowing that SOPs exist for each work activity and that no applicable work activity is to be accomplished without using the proper SOP.

New employee safety and health training for work in the laboratory might also include specific training on other guidelines described in the following:

- Formaldehyde, 29 CFR 1910.1048
- Regulatory Guide 8.29, Nuclear Regulatory Commission (NRC)
- Biosafety in Microbiological and Biomedical Laboratories, Centers for Disease Control Guidelines, HHS Publication (NIH), 88-8395, 1988
- NIH Guidelines for the Laboratory Use of Chemical Carcinogens (NIH Publication 81-2385, May 1981)
- DOT 126-F/HM-181 Function Specific, General Safety, and Safety Awareness Training

5.3.2 SAFETY AND HEALTH TRAINING FOR ALL LABORATORY WORKERS

While it is absolutely essential to ensure that the environmental laboratory has an effective New Employee Safety Orientation Training Program, it is also important that all laboratory workers are thoroughly trained. Experience has shown that safety requirements change. No laboratory worker could reasonably be expected to perform his/her work continuously in a safe manner unless proper safety training is provided on a continuous basis. It is important to remember that if on-going safety training does little more than reinforce what the workers have already learned, then the training is still worthwhile.

5.3.3 THE TRAINER

Safety and Health training is usually conducted by the organization's safety manager, director, coordinator, administrator, or other designated safety person. This practice is logical, but only if the safety person is capable of training. It is important to remember that when teaching, the instructor must be knowledgeable in the material he/she is presenting, but it is just important to be able to present the information.

This last statement might seem ridiculous to some. However, the environmental laboratory generally consists of individuals who have science backgrounds in chemistry, biology, ecology, and environmental science. Although this is an excellent mixture of science backgrounds, the question is: How many lab technicians/scientists, if any, are teachers?

Note: For those individuals assigned the important task of teaching safety and health in the environmental laboratory (or in any other setting), it is important to possess the expertise, the enthusiasm, and communication skills required to effectively present the material. Few undertakings are less boring, meaningless, and downright frustrating than to have to listen to an "instructor" who does not have content competency, lacks enthusiasm, and cannot communicate. Does this scenario ring a bell with you? If so, enough has been said on this critical topic.

On the other hand, for those readers who have not experienced this dreadful occurrence, be thankful.

5.3.4 TRAINING RECORDKEEPING REQUIREMENTS

As pointed out in previous sections, the importance of employee safety and health training can not be overstated. This is especially the case when one considers the training requirements called for under OSHA's General Industry Standards (29 CFR 1910) and specifically under its Laboratory Standard (29 CFR 1910.1450).

When an organization understands its role and responsibility in providing employee safety and health training, it seems reasonable to assume that this same organization will want to properly document its efforts. However, this is not always the case. The writer has, during various audits, discovered organizations that provide excellent health and safety training programs for their employees (as evidenced by lesson plans and video tape training programs) but cannot show or demonstrate a record showing that this training was presented—to anyone.

When this situation occurs (that is, when training is done but is not documented) you should keep in mind how OSHA and the legal system may view this. For instance, you may want to refer to the saying that goes: You can perform all the safety training that you want to perform, but if you didn't document this training, it was not done (Spellman, 1996). This is certainly the view the regulators and legal system may take.

Safety and health training sessions should begin with taking a written attendance record of those who attend the training. The example shown in Figure 5.2 is a type form that can be utilized for this purpose.

When the training session is completed, the training attendance roster should be collected; the information contained on the roster should then be transferred to the work center personnel training records. After the information is transferred to individual training records, it is important to maintain the attendance roster with its original signatures of those who attended on file. This file is the organization's record (evidence) of employee attendance at specified training sessions. It should be pointed

---

TRAINING ATTENDANCE ROSTER

DATE: ____________

SUBJECT MATTER: LABORATORY SAFETY

In accordance with the recordkeeping and training requirements of the Occupational Safety and Health Standard required training for Laboratory Standard 29 CFR 1910.1450 and Hazard Communication Standard 29 CFR 1910.1200, I have received the required safety training. I have agreed to verify my understanding and training of 1910.1450 and 1910.1200 by signing and dating this form. This training meets requirements of 29 CFR 1910.119.

Printed Name	Signature	Work Center
____________	____________	____________
____________	____________	____________
____________	____________	____________
____________	____________	____________
____________	____________	____________

Figure 5.2. A sample of a training roster that can be used to document attendance. *Note*: the reference to 1910.119 is for Process Safety Management training requirements which may or may not be applicable to a particular work center.

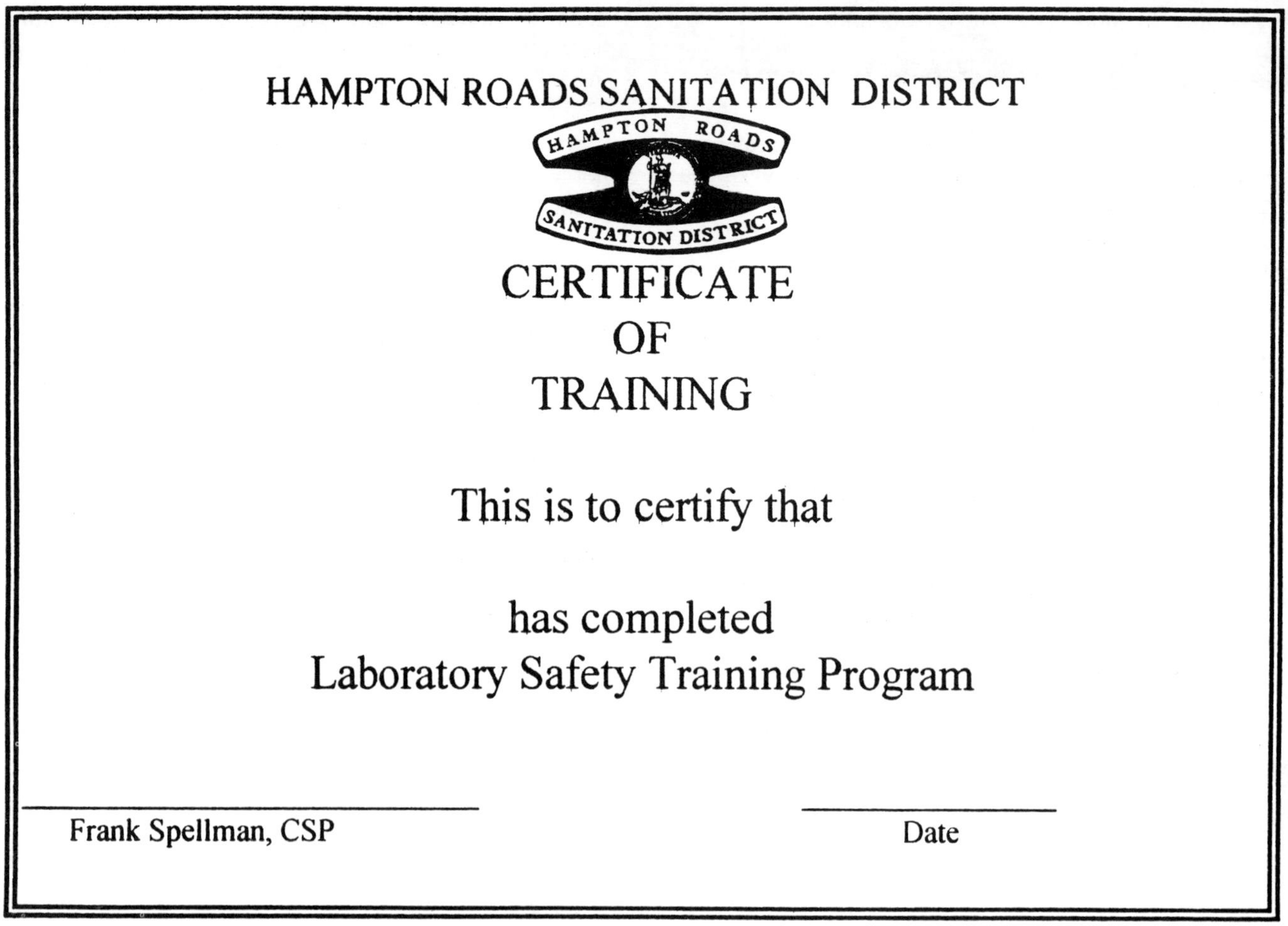
HAMPTON ROADS SANITATION DISTRICT

HAMPTON ROADS SANITATION DISTRICT

CERTIFICATE
OF
TRAINING

This is to certify that

has completed
Laboratory Safety Training Program

Frank Spellman, CSP

Date

Figure 5.3. An example of a type of training certificate that is awarded to Laboratory workers after they complete Laboratory Safety Training. Courtesy of HRSD.

out (emphasized) that this attendance roster is an important document; it will prove its worth when and if it is ever needed.

The work center training records can be of any type—paper, files, books, journals, or computerized—but should include the following information:

- date and time of training
- location of training
- subject matter taught
- description of training
- approximate length of training
- name of attendee(s)
- name of instructor(s)

Along with recording workers' attendance at safety and health training it is also a good idea to present certificates of completion to those who have completed the training. Figure 5.3 shows a sample training certificate that is presented to all workers who attend the training indicated. These certificates can provide other benefits. For example, when a training certificate like the one shown in Figure 5.3 is processed for distribution to the employee, it is prudent to make a copy and place this copy in the employee's individual training record. Experience has shown that outside auditors are usually impressed when they see obvious visual evidence of the organization's training effort—a picture is worth a thousand words.

In summary, training records are not only required they are also important. As a case in point, consider that they provide evidence that the facility acted prudently in providing the worker with instruction that increases the worker's skill and knowledge and informs him/her of the hazards involved, thus reducing the risk to which the worker might be exposed. When you get right down to it, isn't this what safety training should be all about?

CHAPTER 6

Medical Surveillance Requirements

6.1 INTRODUCTION

In its Laboratory Standard, Chemical Hygiene Plan, Respiratory Protection Standard, Hearing Conservation Standard, and Bloodborne Pathogen Standard, OSHA is clear about employer requirements to institute specific medical surveillance guidelines for workers in the workplace. Specifically, for example, the Chemical Hygiene Plan (to be discussed in detail later) requires the employer to institute provisions for medical consultation and medical surveillance for laboratory employees when conditions exist in which exposures in excess of the Permissible Exposure Limits (PELs) or action levels may have occurred, may routinely occur, or have a likelihood of occurring.

OSHA's Respiratory Protection provides another example of when medical surveillance is required. In term's of the user's health, OSHA states that not everyone can wear a respirator. OSHA goes on to state that anyone assigned to work while wearing a respirator must first be cleared by competent medical authority as being fit for respirator use. OSHA also requires followup medical examinations to determine fitness to continue wearing a respirator on the job. These exams should occur at intervals not to exceed 5 years, depending on the worker's age. For workers older than fifty, more frequent examinations are recommended.

Preliminary screening for lab worker suitability to wear respirators can be accomplished by performing spirometry. Spirometry provides a basic assessment of the worker's pulmonary function (suitability to wear a respirator). More will be said about spirometry later in this text.

Another OSHA standard requiring worker medical surveillance is the Hearing Conservation Standard. When a workplace produces noise levels above the **action level** [>85dB(A)], then the workplace must institute a Hearing Conservation Program which also includes medical surveillance. This medical surveillance is in the form of audiometric examinations for all employees who are or could be exposed to noise above the action level. To gain a better understanding of the medical surveillance procedures and training requirements required to comply with the Hearing Conservation Standard, Figures 6.1 through 6.6 are provided.

For formaldehyde exposure or other specific chemical substances regulated by OSHA (for which exposure monitoring and medical surveillance requirements exist in the standard for that substance), the employee must be offered the prescribed medical surveillance program if environmental monitoring shows a routine exposure level above either the action level or Permissible Exposure Limit (PEL), depending on which one is specified.

6.2 COMPLYING WITH MEDICAL SURVEILLANCE REQUIREMENTS

The person responsible for the Chemical Hygiene Plan must ensure that the plan defines the means by which the laboratory will comply with the requirements of the Laboratory Standard. These specific requirements include:

(1) Ensuring that workers who are exposed to chemical hazards have the opportunity to receive medical attention, including any follow-up examination that the examining physician determines to be necessary. For example, if an employee develops any signs or symptoms associated with the chemicals to which he/she may have been exposed in the laboratory, the worker must be provided the opportunity to receive a medical examination. Another example is when an incident occurs such as a chemical spill, leak, or explosion and there is a likelihood (determined via exposure monitoring) that the worker was exposed to a hazardous substance. In this case, the affected worker must

Figure 6.1. Audiometric booth used in conducting hearing tests, which is part of the employers' responsibilities under OSHA's Hearing Conservation Standard.

be referred to medical consultation and, if deemed necessary, medical examination.

(2) OSHA's Laboratory Standard specifically points out that "all medical examinations and consultations shall be performed by or under the direct supervision of a licensed physician and shall be provided without cost to the employee, without loss of pay, and at a reasonable time and place."

(3) OSHA's Laboratory Standard also requires the employer to obtain a written opinion from the attending physician. To facilitate the physician in performing his/her medical analysis the employer must provide the following information, if available:

- the identity of the hazardous substance/material/chemical to which the worker may have been exposed
- quantitative exposure data and a description of the conditions under which the exposure occurred
- information of any signs and/or symptoms of exposure that are experienced by the worker

(4) The employer should receive written opinion from the examining physician. The written opinion should include the following information:

- any recommendations for further medical follow-up
- results of the medical examination and any associated tests
- any medical condition revealed in the course of the examination that may place the laboratory worker at increased risk as a result of exposure to a hazardous chemical present in the laboratory
- a statement that the laboratory worker has been informed of the results of the consultation or medical examination and any medical condition that may require further examination or treatment by the physician

Figure 6.2. Worker being tested in audiometric booth.

Figure 6.3. Just testing a worker's hearing acuity is not enough, according to OSHA. If hearing protection must be worn, then OSHA requires that each worker be given instructions on how to properly use the hearing protection devices. In Figures 6.3–6.6, an industrial hygienist is shown training a worker how to use ear plugs. Along with an explanation on how to wear the hearing protection, the industrial hygienist makes sure that the worker fully understands the organization's Hearing Conservation Program.

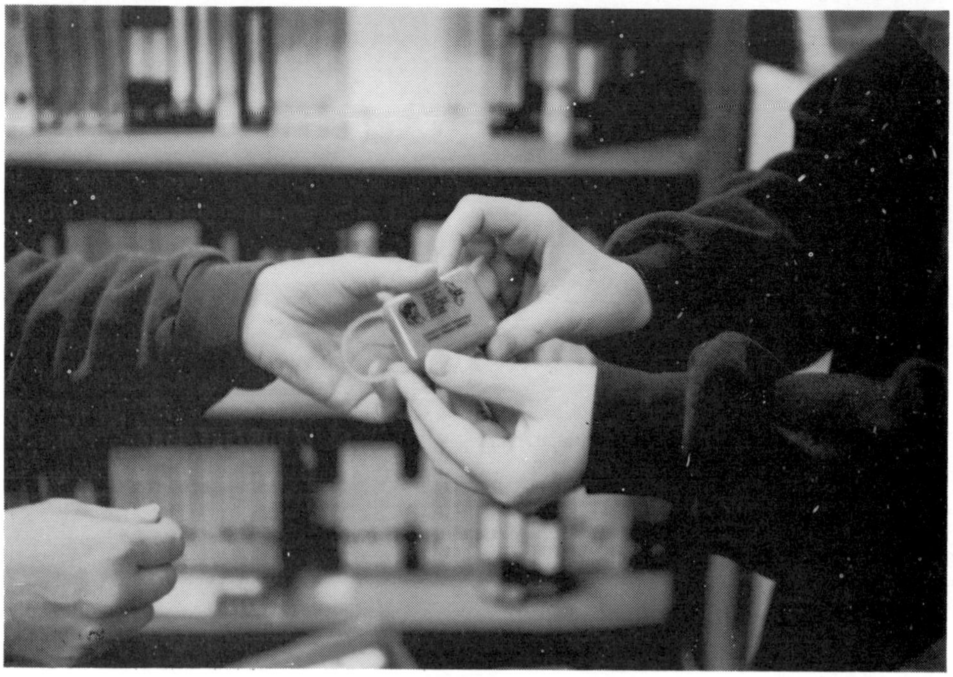

Figure 6.4.

Figure 6.5.

Figure 6.6.

(5) If respirators are used or could be used in the laboratory by the worker, a written medical opinion from a licensed physician must be obtained. This written opinion must state the worker's suitability or unsuitability for respirator use.

Note: Although this requirement is not stated in OSHA's Laboratory Standard, it is stated in its Respiratory Protection Standard; thus, if respirators are used in the laboratory, compliance with this standard is mandatory.

6.3 MEDICAL RECORD KEEPING REQUIREMENTS

Because of the inherent risks associated with working with chemicals, proper record keeping is essential. Moreover, under 29 CFR 1904 and 1910.20, OSHA requires medical records that the employer has acquired for employees pertaining to medical conditions or workplace exposure to maintain and these records and to make them available to employees (and in some cases to OSHA).

Laboratory medical records should be maintained in a confidential file for at least the duration of the worker's employment, plus 30 years. It is important to also point out that OSHA insists that workers have ready access to their medical files. Whenever a worker requests access to his/her record, the facility is required to provide it in a reasonable place, manner, and time (about 14 working days).

CHAPTER 7

Chemical Hygiene Plan

7.1 INTRODUCTION

OSHA'S Laboratory Standard requires laboratory employers to develop and implement a Chemical Hygiene Plan (CHP). This chemical hygiene plan must:

(1) Set forth procedures, personal protective equipment (PPE), safety equipment, and safe work practices that will protect laboratory workers from the health hazards presented by hazardous chemicals used in the laboratory

(2) Keep exposures below the OSHA permissible exposure limits (PELs)

The CHP should include each of the following elements and should indicate specific measures that the employer will take to ensure protection of its laboratory workers.

- standard operating procedures (SOPs) relevant to employee health and safety for work with hazardous chemicals
- chemical hazard control measures that the employer will implement to reduce exposures; that is, control measures such as engineering controls, protective equipment (administrative controls) and hygiene practices
- laboratory ventilation and hoods maintained so that they function properly
- employee information and training provided
- circumstances under which particular lab operations require prior management approval
- provisions for medical consultation and examination
- designation of personnel responsible for implementing the CHP, including a Chemical Hygiene Officer (CHO) and/or a Chemical Hygiene Committee (CHC)
- additional employee protection for work with chemicals that have a high degree of acute toxicity and chemicals of unknown toxicity (e.g., carcinogens, acutely toxic substances, etc)
- provisions for reviewing, evaluating, and updating (if required) at least annually

7.2 CHEMICAL HYGIENE PLAN

It was pointed out earlier that OSHA requires managers of environmental laboratories that use hazardous chemicals to develop a Laboratory Health and Safety Plan. It was also pointed out that a Chemical Hygiene Plan can, under certain conditions, take the place of the lab's Health and Safety Plan. In order to do this the laboratory's Chemical Hygiene Plan must be carefully crafted. This text recommends this practice because a CHP is easier to implement, to maintain, and to update.

The question becomes: What is required to ensure that a facility's Chemical Hygiene Plan meets all the OSHA requirements or guidelines so that it can stand alone as the lab's written safety and health program?

This question is answered as follows: In order to stand alone as ***the*** written laboratory health and safety program the chemical hygiene must include the following elements:

- health and safety standard operating procedures (SOPs)
- criteria for implementing control measures
- requirement that laboratory hoods and other protective equipment (engineering controls) function properly
- provisions for information and training for new and long term employees
- circumstances under which particular operations require prior management approval

- provisions for medical consultations and examinations
- designation of chemical hygiene officer/health and safety officer
- additional protection provisions for work with particularly hazardous substances
- fire protection and prevention
- laboratory safety inspections
- provisions for radioactive materials, infectious agents, and/or controlled substances

It is important to point out that the elements listed above are ***suggested***; that is, they are only suggestions because the OSHA Laboratory Standard is a performance standard. As explained earlier a performance standard allows the employer "license" (a great deal of flexibility) in putting together a particular safety program or plan. Having said this, it is also important to point out that some lessons are learned through experience—such as, the ordeal of having the lab's chemical hygiene plan closely scrutinized during an OSHA audit. From such an experience it is safe to say that if the reader satisfies the elements listed above, he/she will have little trouble with the regulators.

7.3 CHEMICAL HYGIENE PLAN: AN EXAMPLE

Regarding safety and health compliance programs, it is relatively easy for someone to discuss or point out to someone else what is required to be accomplished. The purpose of this text is to do this but also to *show* exactly what is required. Thus, in this section a Chemical Hygiene Plan that has been used for several years and updated regularly is described in full. This particular plan is an excellent example of the type of written CHP that can be used in the environmental laboratory. It is an excellent plan because it has been tested, audited, and corrected (as needed).

HAMPTON ROADS SANITATION DISTRICT CHEMICAL HYGIENE PLAN CENTRAL ENVIRONMENTAL LABORATORY AND SPECIAL PROJECTS LABORATORY (BIOASSAY)

I. INTRODUCTION

Hampton Roads Sanitation District (HRSD) serves thirteen cities and counties in Southeastern Virginia. Moreover, HRSD's 9 wastewater treatment plants process over 225 million gallons of wastewater per day. Currently, one Central Environmental Laboratory performs laboratory services for the plants. Further, the Central Environmental Laboratory (CEL) also services analytical programs including phosphorous removal, municipal storm water management, and specialized metal and organic analyses.

The Occupational Safety and Health Administration's (OSHA's) regulation for "Occupational Exposures to Hazardous Chemicals in Laboratories", 29 CFR 1910.1450 (The Standard), requires that each facility engaged in the laboratory use of hazardous chemicals develop and implement a written program known as a "Chemical Hygiene Plan" which sets forth procedures, equipment, personal protective equipment and safe work practices that will:

- protect employees from the health hazards presented by hazardous chemicals used in that particular workplace
- keep employee exposures to hazardous chemicals below the permissible exposure limits (PELs) specified in the standard

All laboratory personnel are responsible for familiarizing themselves with the District's Chemical Hygiene Plan and adhering to it. Emphasis in the Chemical Hygiene Plan is placed first and foremost on the protecting the employees. Moreover, failure to comply with this requirement is considered cause for disciplinary action. Copies of this written program will be available (for review and reference by all employees) in the following locations within the Central Environmental Laboratory.

1. Outside the Biochemical Oxygen Demand #126 section in the CEL—Right-to-Know Station
2. CEL Library

The basic components of this program are:

1. Safety committee—responsibility
2. Definition of terms
3. Chemical inventory list
4. Basic rules and regulations
5. Material Safety Data Sheets (MSDS)
6. Container labeling
7. Safe work practices
8. Hazardous material awareness, handling and disposal
9. Employee training and information

10. Accident reporting procedure
11. Training records/documentation

II. SAFETY COMMITTEE

The District's Safety Division is the overall coordinator of safety programs and requirements for all departments. In addition, the Safety Division is responsible for organizing, stimulating, and guiding the District's safety programs, as well as keeping up with changes in safety subject areas so as to provide knowledge and information, when required, that is current.

The District's Central Environmental Laboratory (CEL) shall designate in writing a Safety Officer, Assistant Safety Officer, Safety Technician, and Hazardous Waste Coordinator; these names are to be listed in Appendix X of this program. These designated individuals are automatically members of the laboratory's safety committee.

A. Responsibilities

1. Safety Officer
The Safety Officer has the overall responsibility for optimizing safety training, conducting inspections, safety equipment acquisition and updating safety information. If a problem arises, personnel may seek safety & health guidance from the Safety Officer, Assistant Safety Officer, or Safety Technicians. The Safety Officer must see to it that protective equipment is available and used appropriately. The Safety Officer is to make sure that periodic inspections (as specified) of emergency equipment are conducted; equipment such as fire extinguishers, safety showers, fire blanket, eye wash and safety supplies are to be inspected. Further, the Safety Officer is to perform periodic inspections of the laboratory to uncover overlooked hazards, to ensure that lab personnel are following safety rules, and to remind workers to use safe work practices. Any violations of the safety rules will be reported immediately to the Safety Officer or Assistant Safety Officer. The Safety or Assistant Safety Officer will determine if the violation merits disciplinary action; if so, a formal report of the incident will be made in writing to the Chief of the Laboratory Division.

Other Responsibilities include

a. Work with administrators and other employees to develop and implement appropriate chemical hygiene policies and safe work practices.
b. Monitor procurement, use, and disposal of chemicals used in the laboratory.
c. Ensure that appropriate audits of safety equipment, emergency supplies, and first aid supplies are made and that any discrepancies are corrected immediately.
d. Develop safe work practices for new projects in the laboratory and ensure that they are practiced.
e. Know the current legal requirements concerning regulated substances.
f. Seek ways to improve the lab's chemical hygiene program.

2. Assistant Safety Officer Responsibilities

a. Works with Safety Officer, administrators, and other employees to develop and implement appropriate chemical hygiene policies and safe work practices.
b. Assists Safety Officer in monitoring the procurement, use, and disposal of chemicals used in the laboratory.
c. Submits quarterly reports of activities and findings to the Safety Officer.
d. Coordinates and oversees the Chemical Hygiene Policies with the Safety Officer and the Safety Division.

3. Safety Technicians' Responsibilities

a. Make monthly inspections of all equipment such as fire extinguishers, fire blankets, safety showers, and eyewash for proper operation. Submit inspection reports to the Safety Officer or Assistant Safety Officer.
b. Checks monthly to ensure that the emergency supplies are complete. Emergency supplies are to include the following:

1) First aid supplies

2) Acid and base neutralizer powder
3) Spill absorbent powder
4) Mercury spill tamer kit with mercury suction collection bottle
5) Emergency oxygen
6) Fire blanket
7) Dust pan

Reports any deficiencies to the Safety Officer or Assistant Safety Officer immediately.

c. Ensures that protective equipment such as gloves, goggles, laboratory coats or aprons are properly used. Alerts the Safety Officer or Assistant Safety Officer of any infractions.
d. Reports to the Safety Officer or Assistant Safety Officer any safety violations and hazards observed.
e. Reviews MSDS inventory record at least once each quarter.

4. Hazardous Waste Coordinator Responsibilities

a. Maintains records for the types and amounts of hazardous wastes generated.
b. Coordinates with the Technical Services Division for the safe disposal of hazardous wastes by a qualified contractor.
c. Provides technical information to CEL employees regarding special disposal procedures.
d. Ensures that CEL employees are wearing the proper personal protective equipment when disposing of hazardous waste.
e. Ensures that storage containers for hazardous waste meets all OSHA, EPA, and DOT regulations.

5. CEL Building Coordinator's Responsibilities

a. The CEL Building Coordinator will work in concert with the Safety Officer to identify safety hazards and correct them.
b. The Safety Officer will request the Building Coordinator to order and arrange installation of safety equipment and correction of safety hazards.
c. The Building Coordinator will report all safety hazards or unsafe work conditions that come to his/her attention to the Safety Officer or Assistant Safety Officer.

Note: When CEL safety and health issues need to be addressed and decisions made, the above mentioned persons will function as the lab's safety committee to make recommendations to upper management as required.

III. DEFINITION OF TERMS

1. ***Carcinogen***—a chemical is considered to be a carcinogen if:

   a. It has been evaluated by the Internal Agency for Research on Cancer (IARC) and found to be a carcinogen or potential carcinogen; or
   b. It is listed as a carcinogen or potential carcinogen in the Annual Report on Carcinogens published by the National Toxicology Program (NTP); or
   c. It is regulated by OSHA as a carcinogen.

2. ***Chemical***—any element, chemical compound or mixture of elements and/or compounds.
3. ***Chemical Exposure Limit List***—list of some common chemicals stored and used within the District's Central Environmental Laboratory (CEL). The Permissible Exposure Limit (PEL), Threshold Limit Value (TLV), and Immediately Dangerous to Life and Health Limit (IDLHL) are listed where applicable.
4. ***Chemical Manufacturer***—an employer with a workplace where chemical(s) are produced for use or distribution.
5. ***Chemical Name***—is the scientific designations of a chemical in accordance with the nomenclature system developed by the International Union of Pure and Applied Chemistry (IUPAC) or the Chemical Abstracts Service (CAS) rules of nomenclature, or a name which will clearly identify the chemical for

the purpose of conducting a hazard evaluation.

6. ***Combustible Liquid***—any liquid having a flashpoint at or above 100°F (37.8°C), but below 200°F (93.3°C), except any mixture having components with flashpoints of 200°F (93.3°C) or higher, the total volume of which make up 99 percent or more of the total volume of the mixture.
7. ***Common Name***—designation or identification such as a code, name, code number, trade name, brand or generic name used to identify a chemical other than by its chemical name.
8. ***Compressed Gas***—defined as

   a. A gas or mixture of gases having, in a container, an absolute pressure exceeding 40 psi at 70°F (21.1°C); or
   b. A gas or mixture of gases having, in a container, an absolute pressure exceeding 104 psi at 130°F (54.4°C), regardless of the pressure at 70°F (21.1°C); or
   c. A liquid having a vapor pressure exceeding 40 psi at 100°F (37.8°C) as determined by ASTM Method D-323-72.

9. ***Corrosive***—a chemical that causes visible destruction of, or irreversible alterations in, living tissue by chemical action at the site of contact.
10. ***Explosive***—a chemical that causes a sudden, almost instantaneous release of pressure, gas, and heat when subjected to sudden shock, pressure, or high temperature.
11. ***Exposure***—occurs when an employee, in the course of employment, is subjected to a hazardous chemical through any route of entry (inhalation, ingestion, skin contact), and includes potential for exposure.
12. ***Flammable***—describes a chemical that falls into one of the following categories:

   a. Aerosol, flammable—an aerosol that, when tested by the method described in 16 CFR 1500.45, yields a flame projection exceeding 18 inches at full valve opening, or a flashback (a flame extending back to the valve) at any degree of valve opening.
   b. Gas, flammable—1) a gas that, at ambient temperature and pressure, forms a flammable mixture with air at a concentration of thirteen percent by volume or less; or 2) a gas that, at ambient temperature and pressure, forms a range of flammable mixtures with air more than twelve percent of volume, regardless of the lower limit.
   c. Liquid, flammable—any liquid having a flashpoint below 100°F (37.8°C), except any mixture having components with flashpoints of 100°F (37.8°C) or higher, the total of which make up 99 percent of the total volume of the mixture.
   d. Solid, flammable—a solid, other than a blasting agent or explosive as defined in Section 1910.109(a), that is liable to cause fire through friction, absorption of moisture, spontaneous chemical change, or retained heat from manufacturing or processing, or which can be ignited readily and when ignited burns so vigorously and persistently as to create a serious hazard. A chemical shall be considered to be a flammable solid if, when tested by the method described in 16 CFR 1500.44, it ignites and burns with a self-sustained flame at a rate greater than one-tenth of an inch per second along its major axis.

13. ***Flashpoint***—the minimum temperature at which a liquid gives off a vapor in sufficient concentration to ignite.
14. ***Foreseeable emergency***—any potential occurrence such as, but not limited to, equipment failure, rupture of containers, or failure of control equipment which could result in an uncontrolled release of a hazardous chemical into the workplace.
15. ***Hazard warning***—any symbols, pictures, or words or combination thereof appearing on a label or other appropriate form of warning which convey the hazards of the chemical(s) in the container(s).
16. ***Hazardous chemical***—any chemical which is a physical or health hazard.
17. ***Health hazard***—a chemical for which there is statistically significant evidence based on at least one study conducted in accordance with established scientific principles that acute or chronic health effects may occur in exposed employees. The term "health hazard" includes chemicals which are carcinogens, toxic or highly toxic agents, reproductive toxins, irritants, corrosives, sensitizers,

hepatotoxins, naphrotoxins, neurotoxins, agents which act on the hematopoietic system, and agents which damage the lungs, skin, eyes, or mucous membrane.

18. ***Irritant***—a chemical, which is not corrosive, but which causes a reversible inflammatory effect on living tissue by chemical action at the site of contact.
19. ***Label***—any written, printed, or graphic material displayed on or affixed to containers of hazardous chemicals.
20. ***Laboratory***—a facility where the laboratory use of hazardous chemicals occurs. It is a workplace where relatively small quantities of hazardous chemicals are used on a non-production basis.
21. ***Laboratory scale***—is work with substances in which the containers used for reactions, transfers, and other handling of substances are designed to be easily and safely manipulated by one person. Laboratory scale excludes those workplaces whose function is to produce commercial quantities of materials.
22. ***Laboratory-type hood (Enclosed by-pass fume hood)***—a device located in a laboratory that is an enclosure on five sides with a moveable sash or fixed partial enclosure on the remaining side; constructed and maintained to draw air from the laboratory; and allows chemical manipulations to be conducted in the enclosure without insertion of any portion of the employee's body other than hands and arms.
23. ***Laboratory use***—means the handling or use of hazardous chemicals where all of the following conditions are met:
    - Chemical manipulations are performed on a laboratory scale.
    - Multiple chemical procedures or chemicals are used.
    - Procedures are not part of or simulate a production process.
    - Protective laboratory practices and equipment are available and used commonly.
24. ***Material Safety Data Sheet (MSDS)***—is written or printed material concerning a hazardous chemical which is prepared in accordance with approved regulations.
25. ***Mixture***—any combination of two or more chemicals if the combination is not, in whole or part, the result of a chemical reaction.
26. ***Oxidizer***—a chemical other than a blasting agent or explosive that initiates or promotes combustion in other materials thereby causing fire either by itself or through the release of oxygen or other gases.
27. ***Permissible Exposure Limit***—means the maximum time-weighted average (TWA) concentration of a substance in air that an employee can be exposed to legally during an 8-hour shift.
28. ***Physical Hazard***—a chemical for which there is scientifically valid evidence that it is a combustible liquid, a compressed gas, explosive, flammable, an organic peroxide, an oxidizer, pyrophoric, unstable (reactive) or water-reactive.
29. ***Pyrophoric***—a chemical that will ignite spontaneously in air at a temperature of 130°F (54.4°C) or below.
30. ***Reproductive toxins***—chemicals which affect the reproductive capabilities including chromosomal damage (mutations) and affects on fetuses (teratogenesis).
31. ***Threshold Limit Value (TLV)***—a term used by the American Conference of Governmental Industrial Hygienists to express the airborne concentration of a material to which nearly all workers can be exposed day after day without adverse effects.
32. ***Unstable (reactive)***—a chemical which in the pure state, or as produced or transported, will vigorously polymerize, decompose, condense, or will become self-reactive under conditions of shock, pressure or temperature.
33. ***Water-reactive***—a chemical that reacts with water to release a substance that is either flammable or presents a health hazard.
34. ***Work area***—a room or defined space in a workplace where hazardous chemicals are produced or used, and where employees are present.

IV. CHEMICAL INVENTORY

The chemical inventory sheet (along with MSDS) in CEL's Worker Right-to-Know Book lists all the chemicals currently stored or in use. Although it is not stated that all chemicals are hazardous, all chemicals must be handled with CAUTION.

Information regarding the potential hazard of a chemical can be found in the Material Safety Data Sheet (MSDS) for the chemical. Other chemical information sources such as the *Condensed Chemical Dictionary, Merck Index, Hazards in the Chemical Laboratory Manual*, and the *Annual Report of Carcinogens* must also be consulted if other information is required. Consult the Safety Officer or Assistant Safety Officer if information concerning a chemical cannot be found.

V. CHEMICAL HYGIENE AND SAFETY PLAN

A. General Rules

1. Working Alone
Working alone in the laboratory will not be permitted when dealing with hazardous and corrosive chemicals or when the work involves a potentially dangerous procedure. Some areas of work that must be avoided when working alone are:

    a. Analysis involving highly flammable gas such as acetylene, nitrous oxide, methane or propane gas.
    b. Analysis involving the possible poisoning from toxic gas such as the analysis of cyanide or chlorine standard preparation involving bottled chlorine gas.
    c. Preparation of acid or base solutions which involve large amounts of the concentrated solution.

2. Personal Apparel

    a. Shoes must be worn at all times in the laboratory. Sandals, perforated shoes or open- toed shoes will not be worn in the laboratory.
    b. Shorts or torn clothing will not be allowed in laboratories. Use of full-length long sleeve lab coat is mandatory.

3. Personal Housekeeping
All work areas must be kept clean and uncluttered with chemicals and equipment properly labeled and stored. The work area must be cleaned upon completion of a test or at the end of each working day.
4. Horseplay
Avoid practical jokes or other behavior which might confuse, startle, or distract other workers especially in the presence of hazardous chemicals or when a potentially dangerous procedure is involved.
5. Eating/Drinking/Smoking/Applying Cosmetics
Eating and applying cosmetics in areas where potentially hazardous chemicals are present will be allowed. Eating, drinking and applying cosmetics in areas where chemicals are present is a potentially hazardous situation because of the possibility of contaminating the food, drink, or cosmetics with unknown substances. Eating should be restricted to designated break rooms and administrative areas. Drinking will be allowed on a limited basis within the laboratory. Employees may drink soda, or coffee at desk areas, as long as hazardous chemicals are not present and/or not handled. Further, employees must use a closed container such as a soda can, covered cup with straw when drinking within the laboratory. However, food, drink and cosmetics are not allowed on laboratory bench tops. Application of cosmetics is allowed in the administrative area and the restrooms. Smoking in any part of the laboratory (including stockrooms) is strictly prohibited.
6. Unattended Operations
There will not be any unattended procedures involving hazardous substances or dangerous steps taken without prior approval from the Safety Officer. Appropriate warning signs outside areas where hazardous procedures are in progress must be clearly posted. All utility devices such as cooling water (secure connections), gases, etc., must be checked and secured (e.g., compressed gas cylinders 150 pounds and under are affixed to wall/ floor or bench).
7. Glassware
Dispose of all broken or cracked glassware and used syringe needles in proper broken glass and sharps disposal receptacles immediately. Chipped glassware may still be used if it can be fire polished to eliminate the sharp edges.
Use caution when making rubber-to-glass connections. Support lengths of glass tubing while they are being inserted into rubber. Fire polish ends of glass tubing

until smooth. Use a lubricant such as water—**NEVER USE GREASE OR OIL.** Use gloves when making such connections and hold tubing as close as possible to the end being inserted to prevent bending and breaking. Never try to force rubber tubing or stoppers from glassware; cut rubber as necessary to remove.

8. Disposal of Chemicals
No chemical or substance (except water/wastewater between pH 2 and 12.5) should ever be disposed of down the sink/drain without **first** consulting the Hazardous Waste Coordinator or Safety Officer and reading the MSDS.

B. Laboratory Standard Operating Procedures (SOPs)

Every Standard Operating Procedure in the laboratory requires precautionary measures that must be taken in order to execute each procedure safely. The operator must read and understand the procedure thoroughly (especially procedures involving toxic substances) before proceeding. Any questions regarding the safe execution of the procedure *must* be addressed before attempting to execute the procedure. Adherence to all safety work practices outlined below is a *must* to assure safety in the laboratory.

C. Use of Protective Equipment and Apparel

1. Hoods and Ventilation

a. Any operation that might result in the release of toxic chemical vapor or dust must be performed under a properly functioning hood. Use a hood or other form of local ventilation when working with any appreciably volatile substances.
b. Confirm adequate hood performance before each use.
c. Keep the hood ***on*** at all times when it is not in active use if toxic substances are stored in it. However, acid, caustic, and flammable cabinets should be used as much as possible for storage.
d. Keep materials stored in hoods to a minimum and do not allow them to block air flow.
e. When a hood is not in use, keep the sash in the down position.

Confirm adequate hood performance before each use by referring to the air flow monitor on the side of the hood. If the hood does not have a built-in airflow monitor, use a hood air flow measuring device such as a velometer (available in the CEL and District Safety Office). Confirm hood performance with the sash in the down position. Hood face velocity must not be excessively turbulent. Typically, a hood face velocity of 60–100 linear feet per minute is adequate. Also check to see if the hood inspection sticker is current (inspected within the last three months). Do not use a malfunctioning hood and report any deficiency immediately to the Safety Officer or a Supervisor.

2. Eye Protection
The Objective of the lab's Eye Protection Policy is to protect the eyes of CEL employees and CEL visitors. Moreover, the policy pertains to all District employees, vender representatives (demonstrating, modifying, or repairing analytical or analytically-related equipment in the CEL), and all CEL personnel who closely observe analytical activities.

All personnel in the CEL must wear CEL-approved protective eyewear whenever they are involved in the following activities:

- any aspect of analyses except data manipulations
- labware preparation
- solution and sample preparation
- sample handling
- any close observation of an analytical process that requires protective eyewear

There are two general categories of lab situations in which different levels of eye protection are required. These categories are as follows:

General Protection: This is characterized by the in-volvement in analyses or support activities that require the use of materials, samples, or solutions that are not known to result in bodily damage upon contact but could be harmful if eye contact was made. These materials are further characterized as having a pH > 2.0 and <12.0, and a normality <2.0.

Enhanced Protection: This is a situation in which the analysis or support activity involves the use of a distinctly hazardous material. These materials are characterized with a pH <2.0 or >12.0, and/or a normality >2.0. This level of protection is also required when the nature of materials is unknown.

The following types of eyewear are approved for use according to the applicable protection category:

General Protection:
—Prescription glasses with shatter-proof lenses, equipped with side shields. The side shields will be provided by the CEL.
—General safety glasses that comply with the safety standard ANSI Z87.1-1989. These are supplied by the CEL.
—General purpose face shields. These are also provided by the CEL.

Enhanced Protection:
—Safety goggles, chemical splash and impact resistant, indirectly vented; in compliance with safety standard ANSI Z87.1-1989. These are supplied by the CEL.

When contact lenses are to be worn in the laboratory, protective chemical goggles providing protection from flying objects and chemical splash must be worn at all times when performing any type of analytical procedure, with the exception of data manipulations. Any CEL personnel wearing contact lenses, and engaged in any aspect of analysis, must notify their supervisor of this vision correction method prior to analytical activity.

Contact lenses do not provide protection. The capillary space between the contact lenses and the cornea may trap any material present on the surface of the eye. Corrosive chemicals trapped in this space cannot be washed off the surface of the cornea. If the chemical in the eye is painful or the contact lens is displaced, muscle spasms will make it very difficult, if not impossible, to remove the lens.

Certain analytical areas and functions are currently considered not hazardous to eyes. An attempt will be made to accommodate these areas and associated personnel by exempting them from compliance with this policy as long as no obvious eye hazard exists. These areas include the following:

- microbiology rooms
- operation of a computer associated with an analytical instrument (see Figure 7.1)

The success of the exemptions listed above depends on personnel and their ability to routinely remember to wear eye protection in the other areas and for other functions that require the protection.

3. **Body Protection**
Body protecting garments such as long length, long sleeved laboratory coats, must be worn at all times when performing any type of analytical procedure in the laboratory. Laboratory aprons can be worn under unusual circumstances, for example when the air conditioning is out of service. Laboratory aprons can also be worn over laboratory coats for added protection. Body protection will protect the body and clothing from chemical spills such as acids and bases.

4. **Hand Protection (Rubber, Latex gloves)**
The danger of contact with corrosive or toxic chemicals must be minimized by the use of protective latex or rubber gloves. The following are guidelines to follow when selecting gloves:

   a. **Long Length**—gloves are heavy duty, usually green or black, made from a nitrile base, have a 22 ml thickness, and are approximately 18–24 inches long. These gloves are used for working in acid baths because protection is provided to the upper arms and the gloves are acid resistant. Due to the thickness of the glove, better protection against rips and tears which may be caused by broken glass is provided.

Figure 7.1. An environmental laboratory lab specialist using computer to conduct analytical testing procedures.

Note: These gloves can be used when working in *any* acid bath at *any* concentration. Further, these are the only gloves acceptable for working in an acid bath.

b. **Playtex Gloves**—are a "medium weight" glove. The gloves are generally blue or yellow, have a thickness of 25 mls or greater, are made of natural latex, acid resistant, and resistant to rips and tears caused by broken glass. Since Playtex gloves are shorter than the long length gloves, Playtex gloves are not acceptable for use in acid baths. Use Playtex gloves for the following:
   - Working with *any* acid *not* in an acid bath
   - Washing or rinsing of any glassware
   - Loading and unloading the glassware washers
   - Handling any soaps

c. **Common Use Latex/Nitrile Gloves**—Latex gloves are generally opaque-white in color. Nitrile gloves may come in a variety of colors (blue, green, etc.). Both these type gloves are only 4–9 ml thick and only about 9.5 inches long. This type of glove is also acid permeable. Due to thinness, lack of arm protection, and acid permeability, these gloves would not protect the hand or arm against broken glass or chemical spills. These gloves are intended to be used when handling biological samples, such as raw wastewater or biosolids. *These gloves are not acceptable for use in* *the glassware section of the CEL.* Gloves providing thermal insulation, must be used at all times when transferring objects in and out of an oven or furnace.

The type of glove worn should be appropriate to the analysis being performed; when in doubt, consult the Safety Officer.

5. **Chemical Storage and Handling**
Many chemical substances used in a laboratory are toxic to humans when ingested or inhaled. Precautions must be taken with these substances to keep them out of the mouth, nose, eyes, and the skin. It is important to wear protective clothing, gloves, and safety glasses when handling toxic or corrosive chemicals.

The following are specific laboratory rules and precautions that must be followed:

A. Chemical Storage

a. Never place beakers or unstoppered (uncapped) vessels containing chemicals in a refrigerator. Volatile-containing samples should be stored in a special refrigerator designed to accommodate such materials, where available. These are explosion-proof refrigerators especially suited to this application.

b. Label all chemical containers correctly and clearly. The following information must be included in the chemical label:

   1) Date the chemical was prepared with initial of analyst who prepared the chemical.
   2) Date the chemical was received.
   3) Date the chemical was opened and expiration date.
   4) Appropriate hazard warning such as corrosive, flammable, carcinogen, or oxidizer (where applicable).

c. Store acids and bases separately in well ventilated areas and away from volatile organics and oxidizable materials. Do not store strong acids or bases on high shelves.

d. Store large reagent containers in the vented cabinets under the hoods. Ensure that flammable materials are stored in the flammable cabinets. Make sure that all containers used in the preparation of reagents or containers used to hold small amounts of reagents for short periods of time are clearly labeled.

e. All cyanide (CN-) compounds, Barbituric Acid and other regulated substances must be kept locked and accessible only to designated supervisors. Records must be kept of the usage and must contain the following information:

   1) Identity and amount of chemical taken.
   2) Date the chemical was taken.
   3) Initial of the analyst using the chemical.
   4) Intended use of the chemical (type of analysis).

B. Chemical Handling

a. Use fume hoods for all operations involving poisonous or offensive gases or fumes, as well as for operations involving highly flammable or potentially explosive materials.

b. Never heat an organic solvent in an open vessel over an open flame and keep open containers of organic solvents away from open flames or sources of sparks.

c. Avoid pointing the mouth of a vessel being heated towards any person, including yourself.

d. Never add anything to a concentrated acid, base, or strong oxidant; instead, add the acid, base or oxidant slowly and cautiously to the other ingredients, preferably no faster than it is consumed by reaction.

e. Never add solids (boiling chips, charcoal, etc.) to a hot liquid as the result may be violent boiling if the liquid happens to be superheated. Perform such additions when the liquid is still at room temperature.

f. Never pipette anything by mouth. Use an automatic pipettor or a conventional pipette with a rubber bulb.

g. Never pour anything back into a reagent bottle. Consult the Hazardous Waste Coordinator for the proper disposal of unused or contaminated chemical.

h. Prevent accidents such as splashes and container spills. Use a funnel if pouring strong acids, bases or oxidants into a narrow-mouth vessel.
i. Wipe up small spills and bottle rings immediately. A large spill should be handled as indicated in Section 6 "Chemical Spills".
j. Do not flush cyanide compounds down a drain because of the danger of contact with acids flushed down somewhere else and the consequent evolution of toxic *hydrogen cyanide* gas.

Note: Other chemicals due to their high toxicity must be handled with caution. For information regarding the hazard that the chemical might pose, consult the Material Safety Data Sheet (MSDS) or other chemical reference materials discussed in this program. All necessary precautions outlined, such as the use of protective goggles, gloves, body protection, and fume hood must be followed when handling the chemical. Consult the Safety Officer or Assistant Safety Officer if further assistance is needed. Some examples of these chemicals are:

- concentrated acids (hydrochloric, sulfuric, nitric, chromic, phosphoric, glacial acetic)
- concentrated bases (sodium, potassium, ammonium hydroxides)
- cyanide salts and phenols
- mercury and mercuric salts
- heavy metal compounds (Ba, Be, Pb, Cu, Ag, Zn, Cd, Cr, Ni, As, and others)
- oxidizing salts ($AgNO_3$, chromates) (Potassium Dichromates)
- benzene

C. Microbiological Hazards

Potential dangers associated with this type of hazard involves hand-mouth contact while handling contaminated samples or laboratory equipment. Some safety rules that must be used are as follows:

a. Always use protective gloves when handling microbiological standards and samples.
b. Never pipette anything by mouth. Use an automatic pipettor or a conventional pipette with a rubber bulb.
c. Keep contaminated items (cultures, serological discards, etc.) in a separate waste container before sterilizing. Do *not* throw these in with the routine trash.
d. Sterilize or disinfect contaminated materials (cultures, used glassware, serological discards, etc.) by autoclaving before throwing them away or processing for reuse. Use a germicidal soap and hot water, or an autoclave, as appropriate for the type of analysis involved. All cell culture glassware is to be autoclaved.
e. Quickly clean up any spillage. Disinfect the area with bactericide, alcohol, or other type of disinfectant.
f. Frequently disinfect working surfaces with bactericide or germicide. Disinfect hands with dilute germicidal soap.

6. **Chemical Spills**

A. Small chemical spills and leaks must be cleaned up immediately to prevent a potential hazardous situation.

Spill Tamer Kits are provided and located in the CEL as follows:

- Mercury and Acid Spill Tamer Kits located outside Wet Chemistry Lab; these kits contain the following items:

a. Neutralizer solution for acids
b. Neutralizer solution for alkalies
c. Mercury Spill Tamer Solution
d. Mercury Collector Bottle
e. Spill absorbent powder
f. Emergency Oxygen

- Acid Spill Tamer Kit located between Metals Prep Lab and Metals ICP-AA Lab.

B. Chemical Spill Control

Spill Control Measures for Acid or Base Solutions:

a. Concentrated acid spills should be cleaned up as follows: wearing gloves and goggles, add dry acid neutralizer to the liquid spill in excess. Cover all the affected areas repeatedly until all liquid is absorbed. Sweep/shovel the neutralized waste into a trash container. Wash down the cleaned area with copious amounts of water. Neutralize further if necessary, using sodium bicarbonate (baking soda), or commercial neutralizer. Sweep the neutralized wastewater from the spill into the floor drain or use absorbent toweling. It is important to neutralize the acid (or base) rather than trying to relocate in prior to clean-up.

b. Solutions should be contained and absorbed using the Spill Tamer Absorbent from the Spill Tamer Kit.
c. Add dry neutralizer for acids or neutralizer for alkalies until all reaction stops.
d. Sweep up the absorbent and dispose of properly. Consult the Safety Officer or Assistant Safety Officer for proper disposal procedure.
e. Rinse the spill area thoroughly with water.
f. Spills adjacent to a sink which can be washed into the sink safely should be handled in this manner. The area can be neutralized using cold water and dry neutralizer for acids or neutralizer for alkalies, or pH 6.9 solution.

C. Dry Alkali Spills

a. Dry, solids material should be swept together, using the dustpan and brush from the Spill Tamer Kit.
b. Dissolve the alkali in a beaker of water.
c. Flush the alkaline solution into the sink drain with generous amounts of cold water.
d. Rinse the spill area thoroughly with water, pH 6.9 solution, then water.

D. Mercury Spill

Any spill of mercury from a broken thermometer, manometer, etc., can create a serious toxicity problem if cleanup is not handled properly. The mercury will continue to vaporize causing a chronic exposure situation. Mercury poisoning can occur from chronic exposures as well as from a one time exposure to large quantities of the metal.

a. Notify other lab personnel of the spill.
b. Wearing rubber gloves, use the Mercury Spill Kit located in the Spill Tamer Kit.
c. Collect as much mercury as possible, using the Suction Collection Bottle.
d. Sprinkle the area with Mercury Tamer, sweeping to mix the Tamer thoroughly with the remaining mercury. Sweep this mixture up and collect the mixture into a glass or plastic container which can be sealed.
e. Sprinkle the area again with Mercury Tamer leaving it on for several days if possible. (The sulfur in the Mercury Tamer will deactivate the mercury).
f. Dispose of the Mercury Collection Bottle, and other equipment used in contact with the spill, by approved mercury disposal i.e., contain equipment in secure, labeled bag and place in the solid mercury waste disposal drum (Hazardous Waste Storage Area). Dispose of in the solid mercury waste drum.

7. First Aid

A. First aid kits for minor injuries are located in the following areas in the CEL:

- glassware prep area
- biochemical Oxygen Demand Lab
- metals prep lab

B. Burn Emergency Kits for minor burns are located in the following areas in the CEL:

- total solids lab
- metals prep lab

C. Chemical Burns

Emergency burn kits for acid and alkali burns are located in the following areas of the CEL:

- chemical Oxygen Demand Lab
- wet chemistry lab
- auto-nutrients lab
- total solids lab
- metals prep lab
- metals lab

D. Emergency Procedure for Chemical Burns

a. Flush the affected area with copious amounts of water using the emergency shower as needed for at least five minutes. Depending on the situation, get medical care immediately.

b. If first aid directions for burns caused by specific chemicals are available, follow these directions, after the initial flushing with water. First aid directions can usually be found by referring to the Material Safety Data Sheet.

c. *Do not* use oils, fats, salves, or ointments which could assist skin absorption.

E. Chemical Poisoning

Chemical poisoning can result from inhalation, ingestion or absorptions through the skin of a toxic chemical. If you suspect that you or any of your fellow employees have been exposed to any type of chemical poisoning, get help immediately. Proceed as follows:

Determine the source of the poison. It is possible that the whole area might be contaminated with poisonous gas or chemical. If it is determined that the whole area is contaminated, notify all other personnel and have the area evacuated. Secure the contaminated area. Call the poison control center and describe the situation. Do not attempt to enter the area unless you are wearing the proper personal protective equipment and breathing apparatus.

a. Cyanide Poisoning/Inhalation of Poison Cyanide

poisoning can result from ingestion, inhalation, or absorption through the skin. Cyanide can be found in the following forms; inorganic compound (NaCN) sodium cyanide, (HCN) hydrogen cyanide (a toxic gas formed from the acidification of an inorganic cyanide compound), or an organic compound classified as a nitrile. Toxic effects appear immediately with large doses; lesser doses cause dizziness, headaches, or vomiting. Symptoms include cherry-red skin color, rapid breathing, then irregular breathing; collapse, heavy sweating, and convulsions. Rapid attention and treatment is *mandatory*, otherwise death can occur. Because of this, whenever cyanide is being used, notify your supervisor *and* another technician so that help will be rapidly available if needed.

First Aid Procedures for cyanide poisoning:

1. Move the victim to fresh air.
2. Prevent shock by keeping the victim warm.
3. Apply artificial resuscitation if the victim is not breathing. For personal safety do not perform a mouth to mouth resuscitation but instead, use a bag valve mask to revive the victim.
4. Break an amyl nitrite capsule in a gauze pad and hold below victim's nose at chin level for 15–30 seconds out of each minute. Do not administer more than 5 capsules.
5. If the victim has difficulty in breathing, administer emergency oxygen.

b. Ingestion of Chemical Solutions or Poisons

1. Immediately notify another lab technician of the accident and type of material ingested. Obtain medical attention immediately.
2. Do not induce vomiting if the solution swallowed was a strong base (alkali), strong acid, cyanide, strychnine, gasoline, kerosene, any hydrocarbon, or if the victim is having convulsions. Do not give liquids to an unconscious person.
3. For strong acids or bases, give the victim a glass of milk or large amounts of water.
4. Do not give oils, fats, or alcohol unless advised by a physician.
5. Prevent shock by keeping the victim warm.

VI. MATERIAL SAFETY DATA SHEET (MSDS)

The Material Safety Data Sheets (MSDS's) are a set of individual data sheets providing related safety information for each hazardous chemical utilized or produced at the work centers. Material Safety Data Sheets are filed in each work center's "Right-to-Know" book. The "Right-to-Know" books for the CEL are located at the front entrance to the lab and outside the biochemical oxygen demand section.

MSDS's are provided to work centers by Procurement and the Safety Division any time manufacturers forward new copies or new editions.

Work center supervisors are responsible for ensuring that MSDS's are current and available at all times for chemicals at their work center and that chemicals are not used unless this information is available, or familiarity with the chemical is established. Whenever MSDS's are added or deleted, the index of chemicals must be updated.

The Material Safety Data Sheets should contain information as follows:

(1) Identity of the hazardous chemical
(2) Identity of hazardous ingredients in a hazardous chemical mixture
(3) Chemical and physical characteristics of the hazardous chemical
(4) Chemical and physical hazards of the hazardous chemical
(5) Acute and chronic health hazards including signs and symptoms of exposure and medical conditions which are generally aggravated by exposure to the hazardous chemical
(6) Primary route of entry
(7) Personal exposure limits in terms of maximum duration and concentration
(8) Protective measures and special precautions
(9) Emergency procedures and first aid procedures
(10) Date of preparation of the Material Safety Data Sheet
(11) Identification of person or agency responsible for the information contained on the Material Safety Data Sheet

The Material Safety Data Sheets shall not contain any blank spaces. Not applicable or unknown information should be indicated as such.

All chemicals used in the laboratory facility are purchased chemicals. All MSDS are supplied by the chemical manufacturer. Information on the MSDS must be complete and accurate. If the information on the MSDS is not complete or additional information is required, the Safety Officer must be notified before any further use of the chemical. The Safety Officer or Assistant Safety Officer will then contact the chemical manufacturer and request the missing information.

Other information resources are available in the CEL. These resources are located in the CEL Library and are identified in the following:

(1) Condensed chemical dictionary
(2) Compendium of safety data sheets
(3) Hazards in the chemical laboratory
(4) Annual report on carcinogens
(5) Research and industrial chemicals
(6) First aid and personal safety
(7) First aid manual for chemical accidents

For any chemical that exists in the CEL without any available material safety data sheet, the above mentioned references must be consulted to clearly identify the hazard of the chemical, before it is used. If information cannot be obtained, notify the Safety Officer. The chemical will be treated as hazardous and all necessary safety precautions will be taken. If a MSDS cannot be obtained on a chemical and the MSDS is needed for disposal methods or use, the Safety Division will research the chemical and produce a generic MSDS for the chemical. The *CEL Safety Technician* will be responsible for receiving, dating and filing all laboratory MSDS.

VII. CONTAINER LABELING

Chemical container labeling is the primary initial source of warning for employees. This requirement is considered one of the most important sections of this program in ensuring the overall effectiveness of the Chemical Hygiene Plan.

As discussed previously, chemicals used in the CEL are purchased chemicals. Thus, proper labeling of all these chemicals is the responsibility of the chemical manufacturer. Any chemical delivered to the CEL will be inspected upon receipt for proper labeling and completeness of information (MSDS included). If any chemical is not properly labeled and not accompanied by MSDS, the delivery will be rejected or sent back to source of delivery.

The Safety Officer or Safety Technician will oversee this chemical delivery process. CEL personnel who have completed HM-181/126-F training and attended ½ day safety orientation training may take delivery of chemicals. CEL personnel receiving chemicals need to ensure that the following information appears on all incoming chemicals:

- the identity of the chemical
- appropriate hazard warning
- the name of the chemical manufacturer
- the chemical label in good condition and readable
- the chemical container in good condition
- shipping papers in order

If any discrepancy is found on the chemical labeling, packaging or shipping papers, CEL personnel are not to take delivery.

All other small containers or bottles stored in the CEL work area or counter-top MUST be clearly labeled with the applicable information and appropriate hazard warning. If a reagent contains several hazardous materials, the name of the reagent must be clearly labeled on the container. The CEL Safety Officer, Technician or Section Leader will be responsible for ensuring that this procedure is followed.

VIII. HAZARDOUS MATERIAL AWARENESS, HANDLING AND DISPOSAL

A. Administration

The aim of appropriate handling and proper disposal of hazardous materials is to assure minimal harm to people, as well as to the environment. Hazardous waste abatement minimizes the potential for exposure to terrestrial and aquatic life, and it minimizes the cost of clean-up and disposal activities to the District. Awareness is one of the key factors in minimizing exposure to hazardous chemicals. Therefore, all CEL personnel are required to attend ½ day Safety Orientation, HM-181/126-F and Chemical Spill Emergency Response Training.

The Hazardous Waste Coordinator maintains records of types and amounts of hazardous wastes generated and coordinates with Technical Services Division for the safe disposal of wastes by qualified hazardous waste disposal professionals.

B. Identification of the Waste

Each employee must be familiar with the variety of hazardous materials which they might come into contact with in the CEL. For every analytical procedure, one or maybe all of the chemicals involved might be hazardous. Any chemical, even if it poses minimum threat or harm, when not handled safely or correctly can cause severe health damage and/or death. When performing any analytical test, the following procedures must be observed:

1. Familiarize yourself with all the chemicals involved.
2. If information is needed concerning the nature of the chemicals, the employee must consult the MATERIAL SAFETY DATA SHEET (MSDS) or other available reference manual.
3. Consult the Safety Officer of Hazardous Waste Coordinator if necessary.
4. Follow the proper use of protective equipment described in Section V of this program.
5. If the chemical requires special disposal procedures, consult the Hazardous Waste Coordinator.
6. Maintain familiarity with the chemicals as the test may be repeated, or for future reference.

C. Minimizing Hazardous Waste

Minimizing the potential for exposure of humans to hazardous wastes is accomplished by minimizing the wastes.

The volumes of hazardous wastes generated are dependent on the type and number of analyses. Due to these variables, the rate at which waste will accumulate fluctuates.

Constant monitoring of the amount of wastes that are generated in each laboratory section is essential to minimize the waste in the work area. A hazardous waste log book will be established and will contain the following information:

1. Identification and category of the waste. For example, list a waste as cyanide, mercury, acid, and so forth. If the waste contains more than one type of chemical, list all known components of the waste.
2. Amount of waste disposed of.
3. Hazard class (e.g., flammable, corrosive, caustic, acid or suspected carcinogen).
4. Date the waste was placed in the drum.
5. Initial of the analyst who placed the waste into the drum.

In many cases the hazardous nature of the chemical is increased when the chemical is mixed with another type of chemical especially when the chemicals are incompatible. To avoid this situation, chemical waste generated in the CEL will be segregated by type (e.g., acid waste, caustic waste, mercury waste, cyanide waste, and phenol waste).

The Hazardous Waste Coordinator will be in charge of maintaining all hazardous waste log books in the CEL. The Hazardous Waste Log Book will be kept in the hazardous waste storage area. A copy of the log book shall also be kept in the Administrative area of the CEL.

D. Handling and Disposal

1. Handling
Proper handling of chemical waste generated in each lab section will be the responsibility of the analyst at the time the waste is generated. The analyst may dispose of the waste in the hazardous waste drums at anytime; however, they must be authorized to enter the hazardous waste storage area and the two-person rule must be used. The analyst will enter the appropriate information regarding the waste in the hazardous waste log book.
2. Disposal
Containers to be used for storing chemical waste must be the approved type that will not leak or corrode; they must meet compliance requirements of DOT, OSHA, and EPA. Currently, waste is stored in approved drums that are in overpacks to provide double protection. If the waste generated is an unused portion of the chemical, the chemical must be stored in the original container until it is ready for disposal. Consult the Hazardous Waste Coordinator if assistance is needed. All containers used for storing waste must be labeled for their contents. The label shown in Figure 7.2 must be affixed to the waste disposal containers.

Hazardous waste that will stored outside must be protected in such a manner as to prevent their contents from freezing because of the possible danger of the container breaking or splitting when the contents expand. Because there are no electrical outlets in the hazardous waste storage area, the overpacks are wrapped with solar blankets when temperatures drop.

Care must be taken by the employee to use common laboratory safety procedures when handling hazardous materials.

Always be prepared for emergency spills or splattering. Wear appropriate protective equipment. Moreover, when disposing of chemicals wear the following safety equipment:

- goggles
- lab coat
- playtex gloves
- half-mask respirator with appropriate cartridges

Waste will be moved from the general laboratory work area to a central waste storage area at regular intervals to minimize volumes of hazardous materials accumulating in the laboratory. EPA's RCRA regulations state that the waste has to be removed 180 days from the date that the drum was started.

When sufficient quantities of hazardous materials have accumulated, a contractor will be employed by the District to remove the wastes from HRSD property.

The following is a list of hazardous wastes generated by the CEL which will be subject to strict inventory control and will require special disposal procedures.

- phenol
- cyanide salts
- pyridine
- mercury and compounds
- highly concentrated acids and bases
- highly concentrated metal compounds

Every time any of these chemicals is disposed of the amount and identity of the waste *must* be entered into the hazardous waste log book. The hazardous waste log book will be located in the hazardous waste storage area and a copy will be maintained in the administrative area of the CEL. The Hazardous Waste Coordinator is responsible for maintaining the log book that is kept in the administrative area and ensuring that employees are logging in their hazardous waste.

IX. FUME HOOD MANAGEMENT

A. General Information

Fume hoods and other associated protective equipment should be maintained in satisfactory operating condition at all times. Monitoring of performance and any scheduled preventive maintenance should be done in accordance with the manufacturer's suggested guidance. The OSHA Laboratory Standard states that fume hoods need to have a hood velocity of 60 to 100 linear feet per minute to be considered as adequate. In order to insure that lab fume hoods continue to meet this standard the Safety Division performs quarterly fume hood inspections. Moreover,

HAZARDOUS WASTE

FEDERAL LAW PROHIBITS IMPROPER DISPOSAL.

IF FOUND CONTACT THE NEAREST POLICE
PUBLIC SAFETY AUTHORITY OR THE
U.S. ENVIRONMENTAL PROTECTION AGENCY.

GENERATOR INFORMATION:

NAME ______

ADDRESS ______

CITY ______ STATE ______ ZIP ______

EPA ID NO. ______ EPA WASTE NO. ______

ACCUMULATION START DATE ______ MANIFEST DOCUMENT NO. ______

D.O.T. PROPER SHIPPING NAME AND UN OR NA NO. WITH PREFIX

HANDLE WITH CARE!

Figure 7.2. A type of hazardous waste label used to comply with EPA and DOT hazardous waste labelling regulations.

during each inspection fume hood velocity checked. For each hood inspected an individual inspection sheet is used and all entries resulting from the inspection must be made here. Hoods passing inspection receive the sticker shown in Figure 7.3.

Hoods should also be closed at all times except when work is being accomplished or adjustments need to be made. To prevent restriction of air flow, materials stored in the hood should be kept to a minimum and should not block vents or air flow. If at any time an employee feels that the fume hood is not functioning properly he/she should notify the Safety Officer or the Building Coordinator at once.

X. EMPLOYEE TRAINING AND INFORMATION—HAZARD AWARENESS

A. General Information

It will be the responsibility of the Safety Officer/Safety Division to inform employees of the hazards that they may exposed to in the workplace. The primary goal of the Chemical Hygiene Plan is to protect employees from

Figure 7.3. Laboratory fume hood inspection label.

hazards through an extensive training program which will provide the employee the necessary tools required to minimize the risk of exposure in the laboratory.

B. Training and Information Program

In order for employees to be properly informed about hazardous materials in the workplace or specific work area, a training and information program (Safety Meeting) will be conducted on a quarterly basis. Discussion will include the requirements of the Chemical Hygiene Plan and specific problem areas (if any) as well as suggestions from all employees to improve the safety program of the laboratory.
The Safety Officer and/or Assistant Safety Officer will coordinate safety (training) meetings for the laboratory.

C. Key Provisions of Training

1. The requirements of the Chemical Hygiene Plan, its importance and primary goals shall be transmitted to all employees.
2. Information on operations of all work areas where an employee may be exposed to hazardous materials will be discussed in detail. Proper handling and precautionary measures that must be taken to abate the hazard shall be emphasized.
3. The employees shall be informed of the location of the hazard communication materials such as the Material Safety Data Sheets, Compendium of Safety Data Sheets, other safety references and emergency safety supplies.
4. Location and lists of hazardous materials that the CEL generates shall be made available to all employees.
5. Methods and observations to detect the presence of a hazardous material.
6. Use of proper protective equipment.
7. In the event that a new hazardous material is introduced to the workplace, information about the material shall be immediately transmitted to all employees who will be working with that material.
8. Ensure that all lab employees are given training on emergency and evacuation procedures.

D. New Employee Training

All new CEL employees shall attend General Safety Orientation Training given by the Safety Division every three months. While waiting for New Employee Safety Orientation training, all new employees will fundamental safety training which will allow them to start work in the CEL. The new employee shall be indoctrinated on OSHA's Hazard Communication and its Laboratory Standards. Each employee will be informed of his or her role and responsibilities within the program. The employee will be required to sign a form acknowledging that he or she has been informed of the existence of the District's Chemical Hygiene Plan. A record of the employee's acknowledgement shall be kept in the laboratory's training file, maintained by the Safety Division.
All new employees must attend the District's New Employee Safety Orientation. Laboratory employees are required to attend the first 4 hours of the 8-hour orientation session.

E. General Safety Training for All CEL Employees.

All CEL employees shall be trained on the following laboratory safety procedures on a continuing basis.

1. Emergency evacuation
2. Use of PPE 3. Use of fume hood
4. Use of safety eyewash/shower
5. Use of fire blanket
6. Use of emergency oxygen/respirators
7. Emergency response procedures for acid or caustic spills, inhalation or ingestion of poison, medical emergencies, and fire or explosion
8. Use of different types of fire extinguishers
9. Use of the MSDS and other safety reference materials
All safety training shall be documented.

G. Safety Training Videos

Applicable safety films are available to be shown to CEL employees to increase their knowledge of chemical hazards. The Safety Officer can procure films from the District's Safety Video Library located in the Safety Division's Office.

H. Safety Supplies/Equipment

Laboratory safety supplies and equipment such as protective gloves, goggles and aprons or protective apparel, fire extinguishers, emergency eyewash and showers, fire blanket, spill tamer kit, emergency oxygen and first aid kit shall be inspected *EACH MONTH* by the Safety Officer or Safety Technician for completeness and proper operation.

The Safety Officer or Safety Technician will be responsible for keeping safety supplies and equipment in stock and functioning properly.

XI. EMPLOYEE EXPOSURE EVALUATION AND MEDICAL CONSULTATION

This section of the program addresses the need for performing a formal evaluation of a suspected employee exposure to a hazardous chemical in the laboratory.

Contributing factors for exposure to hazardous chemicals such as failure to use proper protective apparel and equipment or failure to follow outlined safe work practices will be evaluated. Any deficiency found shall be corrected immediately to avoid the same problem in the future.

A. Criteria for *Reasonable* Suspicion of Exposure

1. The employee had direct skin or eye contact with the chemical.
2. Manifestation of health hazard symptoms such as headache, rash, nausea, irritation or nose or throat, coughing, dizziness, tearing, irritation or redness of the eyes, etc.
3. Some or all of the symptoms disappear when the employee is away from the work area and into fresh air.
4. Symptoms reappear as soon as the employee returns to the work area.
5. Complaints are received from more than one employee.
6. An unusual odor or smell is noticed in the work area.

B. Exposure Evaluation

The following steps shall be taken to determine the employee's exposure and necessary medical attention needed.

Note: ***If it is necessary for the employee to receive immediate medical attention, appropriate action shall be taken before the following steps are taken.***

1. Interview the employee or the victim initiating the complaint.
2. Gathering of essential information regarding the complaint such as (1) suspected chemical, (2) chemicals present in the work area, (3) symptoms, 4) were control measures such as protective apparel and equipment used; (5) duration of exposure, route of exposure, concentration of the substance in its container; (6) area of the room where exposure occurred; (7) severity and type of symptoms experienced; and (8) are all important factors properly recorded.

3. In cases where there is concern about over exposure to a chemical or chemicals, personal air sampling/environmental air sampling shall be performed by the Safety Division in order to determine the level of exposure.
4. The decision must be made as to the need to send the employee to medical consultation.

A memorandum documenting the complaint and outcome of the investigation regarding the complaint shall be written and forwarded to the Human Resource and Safety Divisions. A copy of the memorandum shall be given to the victim.

C. Medical Consultation

Employees known to have been overexposed to hazardous chemicals shall receive prompt medical attention. The employee will be sent to a medical facility or specialist as assigned by the District. Additionally, a first report of accident must be completed by the employee's supervisor and a copy must be sent to the Human Resource and Safety Divisions. The employee will be notified of the outcome of the medical examination.

D. Exposure/Medical Records

Records of (1) employee exposure to hazardous chemicals, (2) investigation conducted and outcome, (3) medical attention required and (4) results shall be kept on file and made available to affected employees upon their request.

XII. TRAINING RECORDS/DOCUMENTATION

Employee safety training efforts shall be documented (see Table 7.1) in compliance with the Hazard Communication and Laboratory Standards information and training requirements. The documentation shall include the following information:

1. List of employees who attend the training
2. List of absentees to ensure that they receive the required training at a later date
3. Type of training received or list of topics discussed
4. Date
5. Employee acknowledgement

All employee safety training records shall be kept in a binder
titled "Employee Safety Training Records."

XIII. ACCIDENT REPORTING PROCEDURE

A. All accidents and injuries are to be reported immediately.

B. The standard form (generic lab form) will be used to report on-the-job injuries regardless of severity.

C. The supervisor will complete the injury report form with all the necessary information regarding the exact nature of the injury.

D. The supervisor will then submit the report to the Human Resource and Safety Divisions.

7.4 SUMMARY AND MISCELLANEOUS COMPONENTS OF CHP

The Chemical Hygiene Plan for each environmental laboratory will be site specific; thus, it will differ somewhat from sample CHP discussed in this chapter. The importance of the sample CHP discussed here is that it provides a written template upon which you can fashion your own CHP. It should be pointed out that in addition to sample CHP illustrated above there are other component parts that usually make up the laboratory CHP. For example, along with the type of form shown in Table 7.1, your site-specific CHP should contain appendices with the following information:

(1) Phone numbers of poison control center, hospital list, and emergency squads
(2) A copy of the type of First Report of Accident that your firm uses
(3) Physicians list
(4) Designation (by name) of Laboratory Safety Personnel (i.e., Lab Safety Officer, Hazardous Waste Coordinator, Assistant Safety Officer, and Safety Technician
(5) Chemical exposure limit list

TABLE 7.1. Employee Training Record.

TRAINING DATE: ______________________

LIST OF ATTENDEES:	SIGNATURE:
______________________	______________________
______________________	______________________
______________________	______________________
______________________	______________________
______________________	______________________
______________________	______________________
______________________	______________________
______________________	______________________
______________________	______________________
______________________	______________________
______________________	______________________
______________________	______________________
______________________	______________________
______________________	______________________
______________________	______________________
______________________	______________________
______________________	______________________

Topics Discussed/Training Received:

List of Absentees:

PART III

HAZARD IDENTIFICATION AND EVALUATION

CHAPTER 8

Identification of Hazardous Chemicals

8.1 INTRODUCTION

When an environmental laboratory has an effective Chemical Hygiene Plan (CHP) and Hazard Communication (HazCom) Program, two things are almost certain: (1) the hazardous chemicals present and/or commonly used in the work center have been identified, properly labeled, and listed on a chemical inventory sheet; and (2) worker training on the hazardous chemicals used in the work center has been accomplished and proper training records have been maintained.

These two requirements, the proper identification and labeling of hazardous chemicals and worker training on the hazards in the laboratory are required under the various regulations affecting laboratory operations. There is another important area, however, that is sometimes overlooked and not so easy to comply with because it is not always easy to determine: ***chemical exposure evaluation.***

The purpose of this chapter is to discuss the type of chemical exposure evaluation that should be performed in the environmental laboratory.

8.2 IDENTIFICATION OF HAZARDOUS CHEMICALS

Before launching into a discussion of the chemical exposure evaluation techniques that should be employed in the environmental laboratory or in any other facility that uses chemicals, it is important to briefly discuss the identification of hazardous chemicals.

The responsibility for identifying and regulating hazardous chemicals in the United States lies with several regulatory agencies. Obviously, the purpose of regulating the proper identification of hazardous chemicals is to ensure that personnel who are to handle or work with the chemicals are aware of what they are and the hazards they present. For example, the USEPA regulates the use and labeling of pesticides and agricultural chemicals. The U.S. Food and Drug Administration (FDA) regulates the manufacture and use of pharmaceutical products. USDOT regulates interstate transportation and marking of hazardous materials, including those regulated by the EPA and FDA.

The hazardous materials marking system that is regulated by the Department of Transportation employs established guidelines orchestrated by the United Nations (UN). This is the case because hazardous materials are transported all over the world, from one country to another.

8.2.1 UNITED NATIONS/USDOT LABELING SYSTEMS

The UN has developed a two-part labeling/marking system. The first part of the UN numbering system divides hazardous materials into nine hazard classes. *Note*: These nine UN hazard classes have been adopted/adapted, with minor additions/changes, by the USDOT for classifying and marking hazardous materials for transportation (see Table 8.1). The second part identifies transported materials by assigning a four-digit number (which may appear on a standard placard or on a separate orange rectangular panel) to specific materials shipped in tank cars, cargo tanks, and portable tanks (see Table 8.2). It should be pointed out that any facility that routinely produces, uses, receives, handles, and transports hazardous chemicals/materials should keep on hand the latest edition of DOT's *Emergency Response Guide*. This guidebook has become the standard reference for identifying hazardous materials that are shipped by most conventual means of transportation.

Another commonly used way in which to label (and to identify) hazardous chemicals/materials is USDOT's Placarding system. Most people who travel the highways

TABLE 8.1. United Nations/USDOT Hazard Classes.

Hazard Class	Materials	Includes
1	Explosives Class	A,B,C, Blasting Agents
2	Compressed gases	Flammable, non-flammable, poison gases
3	Flammable liquids	Flammable, combustible liquids
4	Flammable solids	Flammable, combustible, water-reactive solids
5	Oxidizers	Oxidizers, peroxides
6	Poisons	Liquids, solids, etiologic agents
7	Radioactive	I, II, III
8	Corrosives	Acids, alkalis
9	Miscellaneous	Hazardous waste

and interstates are familiar with DOT's placards. It is hard to miss these $10\frac{3}{4}$-inch diamond-shaped signs that are fixed to each side and each end of large semi-trucks that travel our roads. Although familiar with these placards, few people know what they really mean. Moreover, not every driver can differentiate between the significance of a placard with UN number 1083 and UN number 1407. However, in some cases the point is made very clear. For example, the skull-and-crossbones symbol as shown in Figure 8.1 sends a pretty strong message; most people have little difficulty understanding its meaning.

Placards displayed on vehicles, tankers, barges, airplanes and railcars are commonplace. As stated above, most people are familiar with these placards and are aware of their intent. It is interesting to note that DOT now requires any person who offers a hazardous chemical/material for transportation to label the package. These DOT placards (labels) for packages are similar to but smaller than the standard placards; they are 4-inch diamonds. These small placards must be fixed to the package in such a way so that they can be readily viewed from any direction (obviously, what this means is that these labels should not be fixed to the bottom of a package only).

8.2.2 NATIONAL FIRE PROTECTION ASSOCIATION (NFPA) LABELING SYSTEM

As stated earlier, the chemical manufacturer is required by various regulations to label the chemical products it produces. Moreover, whenever the manufacturer ships the chemical product by air, water, rail, or road, the shipper must label the product container according to USDOT requirements. Another standard chemical labeling system that is often used in many environmental laboratories is the National Fire Protection Association's (NFPA's) 704 labeling system.

This NFPA 704 labeling system, even though it is basically a voluntary system, has enjoyed wide-range use throughout the U.S. for several years; it has become a nationally recognized method of identifying the presence of hazardous materials. Along with its national appeal, the NFPA 704 labeling system has become a favorite labeling system of local agencies like fire departments, rescue services, and hazardous materials response teams (HazMat Teams). This should not be too difficult to understand especially since, for example, the 704 label installed on the face of a building above a normal entranceway (see Figure 8.2) tells the emergency responders critical information they need to know—instantly. More will be said later about the compelling significance of this powerful symbol.

The NFPA 704 labeling system uses a large diamond shape. The large diamond is divided into four small diamond-shaped sections which are color coded and uses a numerical numbering system for easier identification (see Figure 8.3). The color identifies the kind of hazard and the number identifies the degree of hazard. It should be pointed out that the 704 system identifies the hazard in terms of three categories: ***Health, Flammability, Reactivity (stability-instability)***. Additionally, the severity of the hazard is indicated by a numerical designation from 0 to

TABLE 8.2. Examples of Hazardous Materials by Class and UN Number

Hazardous Material	DOT Hazard Class	UN ID Number
Black powder	1	—
Chlorine	2	1017
Gasoline	3	1203
Hydrochloric acid	8	1789
Phosphorus	4	1381
Styrene	3	2055
Uranium metal	7	2979
Vinyl Chloride	2	1086

Figure 8.1. Skull and crossbones—standard symbol of danger.

Figure 8.2. NFPA-704 label mounted to outside wall of an environmental laboratory. Normally, a label like the one shown here is mounted in close proximity to the entranceway; however, the main point to keep in mind is that it should be mounted where it will be visible to emergency response personnel when they arrive on-scene.

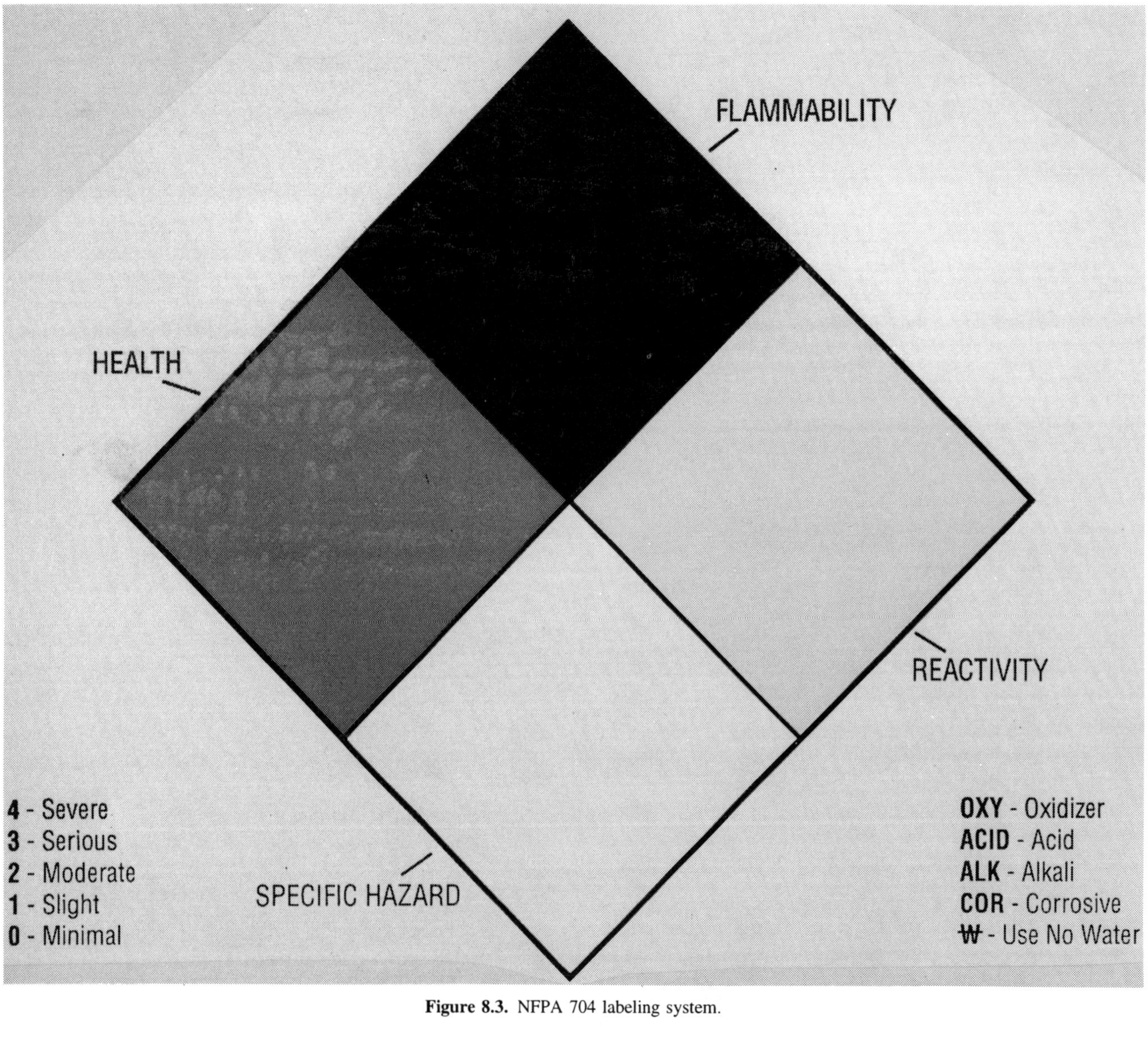

Figure 8.3. NFPA 704 labeling system.

TABLE 8.3. NFPA-704 Color/Numerical Code System.

Hazard	Color Code	Degree of Hazard Rating
Health	Blue	4—Fatal on short exposure 3—Avoid contact/inhalation 2—Harmful if inhaled/absorbed 1—May cause severe irritation 0—No unusual hazard
Flammable	Red	4—Extremely flammable 3—Flashpoint < 100°F 2—Flashpoint 100–200°F 1—Combustible if heated 0—Will not burn
Reactive	Yellow	4—Explosive @ room temp. 3—Explosive if shocked, heated, or mixed with water 2—Unstable, may react with water 1—Unstable if heated 0—Normally stable
Special Hazards	White	W —Avoid water Acid—Acid Alk —Corrosive OXY—Oxidizer —Radiation —others

4. The designations graduate from 0 for a chemical product that has no special hazard, to a 4, which indicates a severe hazard. This color coding scheme and the degree of hazard indicated by each numerical rating is summarized in Table 8.3.

Earlier it was pointed out that most local fire departments, emergency rescue, and HazMat teams have an affinity for the NFPA 704 labeling system. There are several reasons for this: (1) The NFPA diamond is simple; one need not be a brain surgeon to interpret its meaning. (2) The NFPA diamond is user-friendly; it requires no great effort to use. (3) Since the NFPA diamond has wide usage and acceptance throughout the U.S., emergency responders of all descriptions are very familiar with its meaning.

In order to gain better appreciation for the NFPA diamond's powerful, no-nonsense, straightforward meaning, an example of its use is given below.

EXAMPLE 1

At 0315 Monday morning the central fire alarm sounded at fire headquarters in the Township of Rachel's Creek, Virginia. The fire headquarters is also the town's only fire station. From the fire annunciation panel the Fire Captain knew instantly that the alarm was coming from Rachel's Creek Environmental Laboratory. The Fire Captain was somewhat familiar with the environmental laboratory and was well aware that several types of hazardous materials were stored in the lab. Thus, she decided to send both pumper units and the HazMat van to the scene.

With fire sirens piercing the early morning air, the first responders wove their way through the narrow, twisting gravel roads that bisect the township.

Ten minutes later the fire units arrived at the scene. From the beginning, the Fire Captain knew her fire fighters were in for some heavy duty action because the flames, smoke, and heat emanating from the lab building were apparent before they even turned into the parking area in front of the lab.

Fortunately, even though Rachel's Creek is a township of less than 6,200 people, its emergency responders and in particular the Fire Captain responding to this event were well-trained and experienced. In this particular instance, the training and experience paid off.

Upon arrival the fire trucks approached the large one-story brick building with caution. The Fire Captain understood the need for caution—she understood that a mixture of hazardous materials and fire usually brings about catastrophic consequences.

She sat there in the lead fire truck using binoculars to gain a closer look at what awaited them. Above the storefront-like doors fronting the building she spotted what she was looking for: the large NFPA label.

This particular NFPA diamond was labeled as follows: Blue: 1, Red: 1, Yellow: 2. Obviously this information was important; however, the Fire Captain spent little time on this information. Instead, her attention was instantly drawn to the message in the white special hazard diamond: ~~W~~

In the Fire Captain's mind, this particular message lived up to the old adage: a picture (symbol) is worth a billion words!

The Fire Captain knew instantly what she had to do. Actually, more importantly, she instantly knew what she *could not do*! That is, she and her team of fire fighters could not use the only fire extinguishing agent they had available: Water.

Thus, she ordered her responders to retreat to a safe distance where they set up a fire line around the burning structure; they circled the scene with firemen and their fully charged fire hoses. The purpose of this maneuver, of course, was to prevent the fire from spreading.

For the next several hours they manned their positions and watched the lab burn (violently and spectacularly) completely to the ground.

When one takes a look at the event just detailed, he/she might look instantly for fault with the emergency responders. The question might be: Why weren't the firefighters better prepared and equipped to fight a hazmat fire against water-reactive products? This is a logical question, of course. However, in the Rachel's Creek incident just related, the fire department was not prepared to fight a water-reactive type fire because they had no previous knowledge that water-reactive substances were stored in the environmental laboratory.

With whom does the fault lie?

This is the standard question that arises in the aftermath of any disaster. But it is not the point of this example. The point is that, even though this particular laboratory failed to inform the local emergency responders that a recent shipment of more than 250 pounds of lithium had been received and stored in the laboratory, the lab did follow regulatory requirements in labeling the hazard—which directly resulted in warning the responders—*which saved their lives.*

When you get right down to it, isn't this what hazmat labeling is all about?

Example 1 described a situation in which proper labeling afforded the emergency responders the opportunity to make the proper and safe response to a dangerous situation. In Example 2 we look at a situation which is quite different. Example 2 is different in the sense that proper hazardous chemical labeling procedures were not followed and the subsequent results were—to say the least—tragic.

EXAMPLE 2

Dexter's Environmental Laboratory (DEL) had been in business for 8 years. The key words are "had been": Dexter's went bankrupt and out of business a few months ago. The bankruptcy resulted from extraordinary litigation fees brought about by lawsuits generated by former employees and regulatory fines resulting from many instances of noncompliance with Federal, State, and Local Standards. In a few instances, heavy expenditures were proffered for expenses generated both from litigation and fines.

As a case in point, consider the example of Fred Kruger. Fred began work at DEL about five years after the lab opened for business. A high school graduate, Joe was hired as the lab custodian.

The lab manager at DEL, from the first day until closure, was Tom Hardackes. Tom was a college graduate with a Masters in Biological Science. Tom was a difficult person to get to know; he kept to himself. Tom was also driven; he was determined to succeed at whatever enterprise he pursued.

The owner of DEL, Jack Dexter, was like Tom Hardackes in one specific way: he was also determined to succeed. Jack Dexter measured success in terms of how much money he could make and how little of it he had to spend. To say that Dexter was a tightwad would be to make an understatement.

Unfortunately, since the DEL manager, Tom Hardackes, was determined to succeed, he was willing to overlook commonsense, regulatory compliance requirements, including his moral responsibility to provide a safe workplace for his workers. The owner, Jack Dexter, had made it perfectly clear that Tom Hardackes was to find every way possible to save money.

Tom Hardackes knew it was expensive, in regards to employees' time and equipment expense, to support and enforce a strong laboratory safety program. Safety is expensive (no one ever said it wasn't). Moreover, in Tom Hardackes's view, safety was too expensive.

Because he had been directed by the owner to cut costs—to save money anyway possible—Tom Hardackes set out to do just that. DEL operated with as few people as necessary (11 total) and with as little regard to regulatory compliance as possible.

Thus, the stage has now been properly set to relate what all this has to do with Fred Kruger, DEL's custodian.

It all began on that overcast day in early October. One of the laboratory technicians was in the chemical storeroom drawing off a 1-liter sample of concentrated sulfuric acid from a 5-gallon carboy container. After filling the 1-liter bottle, the technician left the chemical room and started back toward his work station.

When he arrived at his work station, a lab assistant met him at the door and stated the lab manager wanted to see him ASAP! Without thinking, the lab technician set the 1-liter bottle of concentrated sulfuric acid on the lab bench and left for the lab manager's office.

The lab manager informed the lab technician that the lab technician had a serious problem at home that needed to be taken care of. The lab technician left for home immediately.

Meanwhile, back in the laboratory, the unlabeled 1-liter bottle of concentrated sulfuric acid sat on the lab bench for the rest of the day.

Each day, a few minutes before closing time, the laboratory assistants routinely cleaned off the lab bench-top work areas. Usually this function was completed with some haste because the lab workers were anxious to head for home.

The unlabeled 1-liter bottle of concentrated sulfuric acid was moved aside toward the back of the workbench. The worker who had moved the bottle had no idea what was in the bottle because the bottle was not labeled.

Later that evening, Fred Kruger was in the DEL to perform his custodial duties. When he came across the unlabeled 1-liter bottle of concentrated sulfuric acid, Fred had no idea what substance was in the bottle. From the outside it appeared to be a pale yellow watery substance. Any other custodian might have left the bottle alone, but Fred was a stickler for perfection. He believed that the laboratory lab tops should be clean and *clear*.

So when he picked up the unlabeled 1-liter bottle of concentrated sulfuric acid his intention was to move it, to place it somewhere other than on the bench top. The problem was he did not know where to put it because he did not know what it was.

Then he made another mistake. Fred picked up the unlabeled 1-liter bottle, unstoppered the bottle top, and brought the container to his nose and inhaled deeply. The horrible acrid vapor attacked his nostrils, throat, and lungs. He jerked spastically, spilling about half the container on his face, throat, and the hand he was holding the unlabeled 1-liter bottle with.

While on his knees screaming in pain, Fred was able to crawl to the nearest emergency shower; however, this was a wasted effort since the shower had been inoperative for several years.

Eventually, several hours later, Fred Kruger was taken to the local hospital where he was treated for severe chemical burns to the left side of his face, neck, and left eye.

For months after this incident, Fred Kruger was the recipient of several sessions of plastic surgery in an attempt to restore his horribly disfigured face. Unfortunately, medical experts were unable to do much with Fred's left eye; it was beyond repair.

Examples 1 and 2 point to the obvious reason why the various regulatory agencies devise and then implement safety and health regulations for industry. It is rather easy to see how proper compliance with OSHA's Laboratory Standard in Example 1 allowed the Fire Department to respond to a serious life-threatening incident in a safe manner. At the same time, the failure to abide by the OSHA Laboratory Standard (Example 2) surely led to the unfortunate accident involving Fred Kruger. In this incident, simple labeling on the 1-liter bottle might have prevented injury to Fred (see Figure 8.4). The proper label could have been the NFPA 704 type or it could been something even simpler. For instance, something simple like a skull and crossbones label, a label containing the word *acid* or *corrosive*, or a label using one of the three standard safety Signal Words: DANGER, WARNING, or CAUTION would have worked to warn Fred Kruger of the danger he held in his hand. The point is that it is doubtful that Fred Kruger would have picked up the 1-liter bottle, unstoppered the cap, and then put the bottle neck to his nose to smell its contents if the bottle had been properly labeled and if he had been properly trained on the label's meaning.

8.3 EVALUATION: EXPOSURE MONITORING REQUIREMENTS

In the preceding section considerable attention focused on the various hazardous chemical identification requirements under OSHA's Laboratory Standard. As with these labeling requirements, the Laboratory Standard contains exposure monitoring provisions that must be complied with. For instance, if the environmental lab uses formaldehyde or other substances specified in substance-specific standards such as OSHA's 1910.1001-1101, it must comply with the exposure monitoring provisions listed in each. Additionally, the lab supervisor must keep in mind that routine monitoring should be conducted where there is reasonable likelihood of worker exposure in excess of allowable levels.

The allowable levels referred to are the permissible exposure limits (PELs) and Action Levels for OSHA-regulated substances. The point is that the environmental laboratory must ensure that employee exposures to hazardous substances do not exceed the specific PELs.

In order to ensure compliance with the various OSHA exposure monitoring requirements, the laboratory should conduct initial monitoring to measure employee exposure to any of the OSHA-regulated substances. If this initial exposure monitoring reveals that employees are being routinely exposed to levels at or above the PEL or Action Levels, then all the exposure monitoring requirements stated in the relevant standards must be observed.

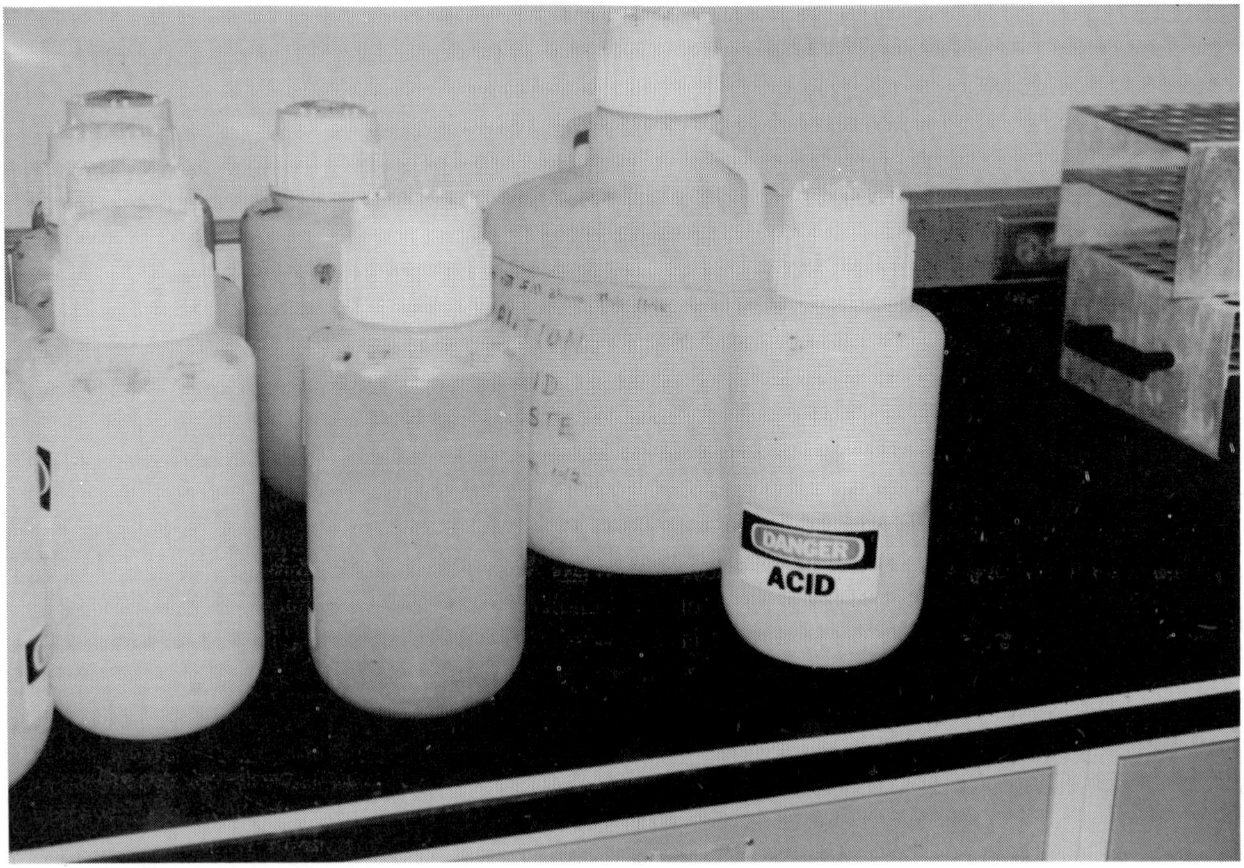

Figure 8.4. Common warning label used in the environmental laboratory (helps to prevent the Fred Kruger syndrome).

8.3.1 TYPES OF CHEMICAL EXPOSURES IN ENVIRONMENTAL LABORATORIES

To this point it has been pointed out that exposures to toxic chemical air contaminants can be harmful and that this is especially the case when the levels exceed permissible exposure limits (PELs) or action levels. Environmental laboratory work typically involves potential exposure to a variety of toxic chemical air contaminants. Moreover, this potential exposure could reach a level that is excessive.

These exposure risks come in a variety air contaminants which may be grouped according to physical state:

- *gases*: Are fluids occupying the entire space of their enclosure and can be liquified only by the combined effects of increased pressure and decreased temperature, e.g., hydrogen sulfide and carbon monoxide. Inhalation is the primary route of exposure to these chemicals
- *vapors*: Result from the volatilization of a liquid or solid chemical at normal conditions or from heating during laboratory operations. Inhalation is the primary route of exposure to vapors
- *particulate matter*: Divided into solids and liquids. In the solid group, there are three categories based on size of the particle and method of monitoring: (1) *smoke*, which is the product of incomplete combustion of organic materials; (2) *dust*, generally created and dispersed by performing a mechanical operation/activity; and (3) *fumes*, which are formed from solid materials by evaporation and condensation
- *mists*: Liquid particles produced by aerosolized mechanical action

8.3.2 LABORATORY AIR MONITORING CONSIDERATIONS

Before beginning exposure monitoring in the environmental laboratory, an assessment of the monitoring considerations should be made. These monitoring considerations have to do with setting priorities. That is, since it is impractical for any laboratory to monitor every square inch of its space, air monitoring priorities must be specified for those workers who may be exposed, on a routine basis, to hazardous chemicals.

TABLE 8.4. Questions That Influence Selection of Monitoring Procedure.

1. How many employees will be exposed?
2. How often and what is the duration of employee exposure?
3. What are the expected airborne concentrations of the chemical?
4. What are the possible routes of employee exposure?
5. What are the chemical, physical, and toxicological properties of the substance to be monitored?
6. What is the representative sample size?
7. Which sampling method should be employed?
8. What impact on sampling will be felt by temperature, humidity, air currents, and other environmental conditions?
9. What range of exposure levels is possible?
10. Will active or passive air monitoring procedures be employed?

Generally, when it comes time to institute air monitoring procedures in the environmental laboratory, a qualitative walkthrough should be conducted. Initially, this walkthrough should be conducted by an experienced industrial hygienist. The industrial hygienist or other professional investigator can accurately evaluate work practices that lead to chemical exposure. In addition, the areas of the lab where personal protective equipment and/or engineering controls should be incorporated to protect the workers can be identified and properly labeled. After the investigator completes the walkthrough and determines the types of operations, chemicals, and number of personnel that could be affected, a sampling strategy should be developed.

8.3.3 FACTORS AFFECTING THE SELECTION OF A SAMPLING STRATEGY

Three major categories of chemical monitoring are usually employed in environmental laboratories: (1) personal monitoring, (2) area monitoring, and (3) wipe sampling. Determining which one of these categories to use depends on the assessments that were made by the investigator during the initial monitoring effort. One thing is certain, in order to perform representative sampling that will relate a true (correct) picture of actual conditions in the laboratory, several questions should be answered prior to selecting a particular monitoring technique. These questions are listed in Table 8.4.

When answers are provided to the questions listed in Table 8.4, then it is likely that the appropriate decision can be made on the type of monitoring method that needs to be employed. It is important to point out that regardless of the monitoring method employed, it is important to collect enough samples to assure that exposures can be accurately determined.

In determining the type of monitoring method to employ there are a couple of key factors to take into consideration. (1) If the goal of the sampling is to measure an employee's inhalation exposure, then the environmental laboratory should use the *breathing zone sampling method*. OSHA defines the breathing zone area for humans as consisting of a hemisphere forward of the shoulders with a radius of about 20 cm. The breathing zone sampling method is considered to be the best method for estimating actual chemical exposure via inhalation. (2) Another factor that is to be taken into consideration is whether it is important or not to measure actual inhalation exposure. If inhalation exposure is not important and area air currents are not a problem, it is common practice to evaluate exposure by measuring work space contamination in the *area* where the workers work. The important point to keep in mind with area monitoring is that it provides a general idea of potential exposures. (3) When attempting to investigate chemical dispersion from an exposure source or in determining whether a worker's protective equipment is free of contamination, *wipe sampling* is often the method employed. This technique utilizes surface area wipes to determine if a particular chemical has been significantly dispersed throughout the work area.

Whatever air monitoring technique is used in sampling in the environmental laboratory it is important to ensure that the sampling results are accurate. In order to make a proper determination of airborne concentrations of contaminants it is essential that (1) the proper sampling equipment is utilized and handled properly; (2) all sampling equipment used is properly calibrated; (3) correct analysis is performed, avoiding analytical and calculation errors; (4) sample contamination is avoided; and (5) results are properly recorded.

8.3.4 RECORDKEEPING AND NOTIFICATION OF MONITORING RESULTS

It is important to note that under OSHA's laboratory standard and its 29 CFR 1910.20 the employer must allow employees access to personal exposure and medical records. More specifically, these regulations mandate that the employer inform workers of monitoring results within 15 working days after their receipt. Usually these results are posted in the laboratory in a location that is accessible to all lab workers. Additionally, individual exposure records must be maintained and made available to any lab worker upon request.

8.3.5 BIOHAZARDS AND RADIATION HAZARDS

The typical environmental laboratory normally performs analytical testing, consulting or monitoring on sick buildings, hazardous waste, lead and paint, compressed air/gas, water/wastewater, petroleum products analysis, stormwater, asbestos and state NPDES Permits. The point is that the typical environmental laboratory is not involved with handling biohazards such as infected body fluids, body tissues, or disease-laden materials. Thus, with the exception of wastewater and/or contaminated water, environmental laboratory workers are not normally exposed to biohazards. Moreover, the typical environmental laboratory does not expose its workers to ionizing radiation; radioactive equipment and materials are not normally used or found in a typical environmental laboratory.

If the environmental laboratory does expose its workers to biohazards (e.g., wastewater lab workers), then the lab must incorporate a biosafety program into the lab's overall safety program. In addition, if radioactive sources used in radioactive test devices are utilized in the lab, the lab must have a safety program that specifically addresses the hazards of ionizing radiation.

For the purposes of this text, it is assumed that the typical environmental laboratory described herein does not expose its workers to biohazards or radiation. Therefore, this text does not specifically address, in detail, biosafety and/or radiation safety.

CHAPTER 9

Waste Management: Hazards and Practices

> As the demand for a clean, safe environment grows, so also the public demand for protection from the horrors of hazardous waste mismanagement. Entrepreneurs of industry and commerce provide daily evidence of the general awakening to the need for reduction or elimination of hazardous waste sources, and better management of the wastes that are generated. But the ever-present drive for new product advantage, competition, budget and capital constraints, and the activities of those who have not yet accepted their environmental responsibilities continue to threaten our environmental well-being. (Blackman, 1993, preface, n.p.)

9.1 INTRODUCTION

ONE point is certain, all laboratory work with chemicals eventually generates some chemical waste, and those who produce such waste have a legal and moral obligation to see that the waste is disposed of in ways that pose minimum potential harm. In the "old" days this meant pouring the chemical waste down the nearest convenient drain. This type of hazardous chemical disposal is no longer tolerated.

Laboratory waste is generated in widely varying operations of different sizes and complexity; producing varying varieties and amounts of wastes accordingly. The employment of a particular waste management system in a specific laboratory setting is, obviously, dependent on the particular laboratory. This is not too difficult to understand when one considers that each laboratory is different in size and in function. However, each laboratory, no matter what type, has basic characteristics/requirements that are common to each. For example, because the laboratory generates waste there must be (1) a commitment of the laboratory manager to the principles and practice of good laboratory management, (2) a written waste management plan (3) assigned responsibility for the waste management program, and (4) policies and practices directed at reducing the volume of waste generated in the laboratory (National Academy of Sciences, 1983).

Along with having basic characteristics/requirements that are common, each laboratory, in its written waste management program, should have two basic goals:

(1) To operate the lab in compliance with all applicable regulations, guidelines, and good industry practices
(2) To manage the wastes generated in a manner that protects workers, the public, and the environment (Stircoff & Walters, 1995)

9.1.1 RESOURCE CONSERVATION AND RECOVERY ACT (RCRA)

In regards to complying with applicable waste management regulations, it should be pointed out that the federal government took the lead in this area with its Resource Conservation and Recovery Act (RCRA) ("Cradle to Grave Act") of 1976, amended since. RCRA's major emphasis is the control of hazardous waste disposal. Specifically, RCRA governs the generation, transportation, storage, treatment, and disposal of hazardous waste. Under RCRA, several responsibilities are specified for generators of wastes. For example, a waste generator must (1) determine if the wastes are regulated as hazardous, (2) determine its generator status, and (3) comply with the regulations specific to its generator status. RCRA is commonly known as the *Cradle to Grave Act* because the EPA (usually via state regulators) assigns cradle-to-grave responsibility for wastes it produces.

9.1.2 COMPLYING WITH RCRA

In order to comply with RCRA, the environmental laboratory must first determine whether the wastes it generates are hazardous wastes. According to RCRA, a waste is defined as any solid, liquid, or gaseous substance that is no longer used and will either be recycled, disposed of, or stored for later treatment or disposal. Laboratory work involving chemicals produces chemical wastes that can potentially cause serious problems if not handled and disposed of properly. The question becomes: Is the laboratory producing or generating waste that can cause injury, death or environmental damage? If the answer to this question is yes, then the waste must be considered hazardous. Thus, this hazardous waste is regulated under RCRA and state public health and safety laws.

A waste can be brought into the hazardous waste regulatory system either by being specifically listed as hazardous (40 CFR 261), or is identified as hazardous through characteristics. The listing of hazardous wastes is accomplished generically under RCRA. The point is that if the EPA has reason to believe that a laboratory waste is hazardous and meets a generic listing, the waste is automatically regulated. If a particular waste does not appear on the EPA lists, it is still to be considered hazardous if it has one or more of certain characteristics. These characteristics are listed as follows:

- *ignitability*: Flash point is below 140°F.
- *reactivity*: It undergoes violent changes, that is, reacts violently with water, explosives, or a sulfide or cyanide-bearing waste.
- *corrosivity*: pH is less than or equal to 2 (acid) or greater than or equal to 12.5 (base).
- *toxicity*: Leachate from the waste is likely to release hazardous constituents into groundwater. The laboratory test for evaluating wastes for toxicity is the toxic characteristics leaching procedure.

It is important for environmental laboratory to take a look at the types of analytical tests it performs and the chemical inventory it maintains to determine the type of waste streams it might generate. Typical environmental laboratory-generated waste streams that are regulated include: acid/bases, ignitable wastes, heavy metals/inorganics, reactives, and solvents. When these waste streams are identified, then it is important to determine which of these wastes are regulated as hazardous under RCRA. It is important to keep in mind that even very small amounts of some waste streams are regulated. These smaller quantity/volume waste streams usually include chemicals, listed in 40 CFR 261, that are "acutely hazardous" because they are identified as being very dangerous.

9.1.3 HAZARDOUS WASTE GENERATORS

All generators are required to manage their hazardous wastes safely. In order to accomplish this, of course, the environmental laboratory must first establish whether it is a generator under RCRA. As soon as the laboratory identifies that it is a generator of hazardous wastes, it must then quantify the amount of the waste generated to determine the regulatory status applicable.

There are three types of hazardous waste generators based on quantity.

- large quantity generator
- small quantity generator
- conditionally exempt small quantity generator

These hazardous waste generators are described in the following:

(1) If an organization generates 1000 kg (2200 lb), or 300 gallons or more, of hazardous waste, or more than 1 kg of acutely hazardous waste in any month, the organization must meet all of the requirements of a ***large*** generator, as specified in 40 CFR 262 or as applicable via modifications instituted by each state.

(2) If and organization generates more than 100 but less than 1000 kg (220 to 2200 lb), or 30 to 300 gallons, of hazardous waste and no more than 1 kg of acutely hazardous waste in any given month, the organization is considered a *small-quantity* generator and is subject to less stringent regulatory requirements than the larger volume generators.

(3) If an organization generates less than 100 kg (220 lb) or 30 gallons of hazardous waste and no more than 1 kg of acutely hazardous waste in any calendar month, it is considered a *conditionally exempt, small-quantity* generator. In order to remain classified as a conditionally exempt small quantity generator the organization is required to:

- Identify all hazardous waste it generates.
- Send waste to a hazardous waste facility, landfill, or other facility approved by the state.
- Maintain accumulations of hazardous waste below the 1000 kg level. If hazardous waste accumulated exceeds the 1000 kg level, it will then be subject to all applicable hazardous waste regulations.

Note: It is important to point out that an organization may be regulated under different rules at different times. This depends, of course, on the amount of hazardous waste the organization generates in a given month. For example, a conditionally exempt small quantity generator will become a small quantity generator if the organization generates between 100 and 1000 kg of hazardous waste in any month.

9.2 WASTE MANAGEMENT PRACTICES

Because of the requirements under the Hazardous and Solid Waste Amendments (HSWA) of 1984, all generators must certify that they have in place a program to reduce the volume or quantity and toxicity of the waste they generate. Moreover, HSWA also mandates that the generator certify that its current method of management is the most practicable method available to minimize present and future threat to public health and the environment.

In line with the requirements stated under HSWA, it is prudent for laboratory management to follow a set of generally accepted waste minimization practices. These practices can work to make compliance with waste management regulations easier. Moreover, using these practices works to reduce waste-related costs and to minimize environmental impact.

When you get right down to it, the heart of effective waste management practice is an effective waste minimization program. This is the case because an effective waste management program cannot concern itself solely with disposal of waste. Disposing hazardous waste is an expensive practice. Moreover, disposing hazardous waste is also wasteful. Chemicals that make up the chemical waste stream begin as raw resources. The point is they are resources that should be conserved and not wasted.

It was pointed out earlier that one of the requirements placed on waste generators is that they must certify (document) that they are making an effort to reduce the amount of waste that they generate. In this waste minimization effort it is important for the lab to develop and implement waste handling and treatment options that are prioritized as listed in the following:

(1) Recycling
(2) Reuse
(3) Recovery
(4) Treatment
(5) Disposal

From this list of priorities it can easily be seen that recycling chemicals is paramount to final disposal.

As pointed out earlier the environmental laboratory should follow generally accepted practices involved with handling and minimizing hazardous waste. The practices pointed out in the following are designed to provide guidance for the implementation and practice of waste minimization.

(1) ***Segregation of wastes***: By not mixing nonhazardous and hazardous wastes, by segregating hazardous wastes, and by avoiding chemical spills or leaks the volume of wastes and the cost to dispose of the wastes can be minimized.

(2) ***Planning analyses***: Each lab task that involves the use of hazardous chemicals should be evaluated to determine what measures can be taken to reduce or eliminate the amount of waste generated.

(3) ***Using smaller quantities in analyses***: With the advancements in technology and especially in computers, it is now possible to reduce quantities of certain hazardous substances used in analyses. Consider, for example, that it is now possible to interface a computer with analyses to record variables and automatically record and interject them when needed (see Figure 9.1). This may allow the analyst the opportunity to make one analysis instead of several.

(4) ***Elimination of orphan wastes***: Orphan wastes are wastes that are of unknown origin and characteristics. These wastes are generally the result of a manufacturer's failure to label or "properly" label a specific hazardous substance. Orphan wastes are also generated whenever lab workers forget to label a hazardous substance and then leave the substance on a lab bench (i.e., the old Fred Kruger scenario) (see Figure 9.2). When orphan wastes are generated, the lab must handle the material as a hazardous waste and properly dispose of it. Obviously, this is the type of work practice, with its generation of hazardous waste, that must be avoided.

(5) ***Substitution/elimination***: Simply stated, substitution is the replacement of a hazardous substance with a non- hazardous substance. An inventory of the toxic or hazardous materials commonly used in the laboratory should be made to determine the possibility of substitution with less toxic or hazardous substances. The next step beyond substitution, of course, is elimination. Each laboratory standard operating procedure should be reviewed and evaluated to determine if alternative methods can be implemented that would eliminate the need for the hazardous substance.

Whether substitution or elimination is incorporated, the point is that not only do these actions reduce or eliminate the amount of hazardous waste generated but they also reduce the health, safety and environmental risks associated with hazardous substances.

(6) ***Purchasing: small versus large quantities***: It is common practice to purchase commodities in large quantities or in volume because this usually results in lower prices—in saving money. Likewise, chemicals sold in large volume or in large containers usually cost less than the same chemicals sold in smaller quantities or containers. On the surface it seems reasonable to assume that a laboratory manager will purchase chemicals for use in the lab at the best price he/she can get (meaning, that large containers costing less will be purchased). However, if this same lab manager were to factor in the cost of disposing of unused chemicals, the purchasing decision might favor the smaller container.

(7) ***Inventory management***: It does not take long, a few years or less, before storage shelves fill up with chemical stock. This is especially the case when the lab

Figure 9.1. Computer station used for laboratory analysis. In this particular environmental laboratory, a similar computer station is installed in most of its 24 individual lab stations.

does not utilize some system for management of the chemical inventory. Exacerbating the problem with overstocked inventories of chemicals is their shelf-life. Because of this shelf-life (usually 1 to 2 years) many chemicals end up occupying shelf space for the entire length of their shelf-life and reach expiration date before ever being used. What this really means is that the chemical stockroom becomes, in many instances, a generator of hazardous waste. This is the case, of course, because when these chemicals and/or hazardous agents expire, go out of specification, or degrade, they must be disposed as hazardous waste.

Not all the news pertaining to chemical inventory control is bad, however. Phifer and McTigue (1990) point out, for example, that "perhaps no other small quantity generators have more opportunity to control their inventory and reduce disposal costs than laboratories. Analytical and bench chemistry procedures developed over the last 20 years have made the use of small quantities of chemicals more practical in the laboratory. . . . The purchase of smaller quantities makes sense" (p. 124).

An important part of good chemical inventory management practice is to follow an inventory routine that includes the following steps:

- A ***first-in/first-out*** policy requires the use of older chemicals before newer ones.
- A ***just-in-time (JIT)*** policy provides for the needed chemical or hazardous agent when needed and in the quantity/volume required.
- When it is necessary to store a chemical or hazardous agent, ensure that it is properly labeled. This label should not only include the chemical's common name but also the the date it was received and its in expiration date.
- A written chemical inventory should be maintained on file or in computerized format.
- To ensure that the chemical inventory is up-to-date and reflects the materials on-site, supplemental inventories should be conducted periodically, and any expired chemicals and hazardous agents identified during the reviews should be removed from the laboratory.
- It is important to document the usage data generated during the inventory to determine the

rate of chemical use and to eliminate, hopefully, unnecessary chemical purchases.

A final word on chemical inventories in the lab: Because chemical reagent distributors sell chemicals in the traditional 500-gram and liter quantities, an overwhelming percentage of laboratory waste disposed of off-site is unused or outdated chemicals that were purchased in excessive quantities. Clearly, these chemical distributors need to get on the environmental bandwagon and live up to RCRA's "less is better" philosophy. This philosophy not only reduces the amount of waste to be disposed but also, correspondingly, reduces the costs of laboratory waste disposal (National Academy of Sciences, 1983). Shouldn't this be the goal in laboratory inventory control?

(8) ***Reusing chemicals***: Many chemicals that have reached their expiration dates may still have some practical value. For example, expired chemicals that are still stable and not degraded may be used for on-the-job training purposes. Some large environmental labs are not only well-staffed but also well-equipped. These labs may have the expertise and equipment available to reuse, for example, contaminated solvents, spent mercury, and silver from photographic processes. Because some chemicals are more easily recovered than others it may be beneficial to review the laboratory chemical inventory to determine if substitute chemicals that are more readily recoverable can be used.

9.3 ON-SITE HAZARDOUS WASTE STORAGE

No matter which hazardous waste quantity/volume reduction plan or reuse technique is utilized in the environmental laboratory, the fact of the matter is that some hazardous waste will be generated. Thus, it is important for lab management to take certain steps to ensure that proper hazardous waste storage is (1) provided; (2) is properly equipped (i.e., containment, heat to prevent freezing, and spill kits nearby); and (3) on-site hazardous

Figure 9.2. The importance of container labeling cannot be overstated.

Figure 9.3. A hazardous waste storage area. Notice the way in which it is labeled and secured.

waste is stored in suitable containers for the type of waste generated (see Figure 9.3).

Before actually providing a space or facility within the environmental laboratory for waste storage, it would be prudent to review the pertinent regulations for on-site hazardous waste storage to determine exactly what is required of the storage facility and how it is to be maintained.

Along with local regulations, environmental labs generating small quantities of hazardous waste also must be aware of federal on-site storage regulations for small quantity generators. For example, under RCRA, the small quantity generator may store no more that 6000 kg of hazardous waste on site for up to 180 days, or for up to 270 days if the waste must be shipped to a treatment, storage, or disposal facility which is located over 200 miles aways. Moreover, small quantity generators are allowed to store for as long as 180 to 270 days so that they will have time to accumulate enough hazardous waste to be shipped off-site. It is important to keep in mind that if a small quantity generator exceeds either the quantity or time limit, the lab will be considered a storage facility and it must obtain a storage permit and meet all of the RCRA storage requirements. The time limit for full-compliance generators is 90 days.

Hazardous waste should not be stored like other wastes. In other words, you wouldn't want to store a 55-gallon drum of phenol waste in the same manner that paper waste would be temporarily stored. Instead, hazardous waste should be stored in an isolated, dedicated, well-labeled area (see Figure 9.3). This storage area should be equipped with containment that will contain any hazardous waste spill that occurs within the storage area. This storage area should be equipped with proper lighting, ventilation, and heating. Heating is important, but often overlooked. For example, whenever a 55-gallon drum of hazardous waste is stored in an outside storage area in locations where winter can be harsh (temperatures below freezing), there is the possibility that the 55-gallon drum might burst and spill its contents if it freezes. It is also important to remember to locate emergency response/spill equipment in close proximity to the hazardous waste storage area. These spill kits help you respond quickly to clean up the spills encountered. More will be said on emergency spill response later in this text.

A final but important point to remember about hazardous waste storage is that the containers used to store the waste should be suitable for the type of waste generated. Hazardous waste storage in unauthorized containers can lead to leaks or accidental spills that could impact human health and the environment; thus, we are talking about an important safe work procedures and safe work practices for the laboratory that must be practiced.

In regards to recommended safe work practices that should be used with on-site hazardous waste storage procedures, the following guidelines for on-site hazardous waste storage in containers is provided:

(1) Clearly label each container with the words HAZARDOUS WASTE—on the label mark the date the waste begins to accumulate.
(2) Inspect each container to ensure that it is in good condition and that each is without leaks.
(3) Ensure that containers used are corrosion-, leak-and rupture-resistant types.
(4) Except for filling or emptying, ensure that hazardous waste containers are kept closed/sealed at all times.
(5) Inspect, periodically, each container for corrosion, leaks, rupture, or other failure.
(6) Ensure that the laboratory's buffer zone between the facility and the property line is maintained by storing reactive or ignitable waste as far from the property line as possible.
(7) Absolutely ensure that incompatible wastes are not stored in the same container.
(8) Remember to ship stored wastes off-site or treat on-site within 180 or 270 days.

9.4 TRANSPORTING HAZARDOUS LAB WASTE OFF-SITE

Most laboratories contract the services of a hazardous waste disposal company to dispose of hazardous wastes off-site. Since the ultimate responsibility for damages caused by a hazardous waste is borne by the generator, it is crucial that a reputable and licensed disposal company/contractor be used. These hazardous waste disposal contractors normally dispose the waste in an EPA-permitted hazardous waste landfill. Most disposal contractors will accept a variety of wastes including solids, liquids, and lab packs. These wastes must be put into containers acceptable for transportation prior to shipment and must be clearly and properly labeled.

Probably the most common type of laboratory container used in disposal of laboratory waste is the *lab pack*. A lab pack is an overpack container (usually a steel or fiber drum of 55-gal capacity with removable lid) that is filled with small quantities of chemicals of the same hazard class packed inside with an inert packing material capable of cushioning bottles and absorbing spills from any liquid containers in the drum. Lab packs are sent off-site for processing, usually either by burial in a secure chemical landfill or by incineration. The maximum volume of waste contained in the lab pack is about 15 gallons.

9.5 SHIPPING DOCUMENTS

In order to ship hazardous waste from the laboratory to the disposal site the proper shipping document must be used. This shipping document is called a Uniform Hazardous Waste Manifest. The manifest is a multi-copy document which the generator must fill out and use to accompany its hazardous waste shipments. The manifest is a chain-of-custody document that is designed to track waste from point of origin to final destination (RCRA's "Cradle to Grave" system). The manifest is an important document that must be handled with attention to detail and to procedure. For example, the hazardous waste generator, the transporter, and the designated treatment/storage/disposal facility must each sign this document and keep a copy. The treatment/storage/disposal facility operator must also send a copy back to the generator—this assures the generator that his shipment arrived at its destination. The generator is also required to maintain a copy of the manifest, with all appropriate signatures, on file for 3 years.

PART IV

ENGINEERING CONTROLS

CHAPTER 10

Laboratory Design

Safety First and Early in Design

10.1 INTRODUCTION

THE statement heading this chapter, **Safety First and Early in Design** (Bretherick, 1991), seems so fundamental to good common-sense especially relating to construction and safety that it seems strange to this writer to even mention it. Strange, yes, but, unfortunately, quite necessary. Why? Well, for example, experience has shown that in the past designers and builders have designed and built ships that provided little if any provision for abandoning ship in case of emergencies. In this particular instance the fitting statement is: **Design It, Build It, and Forget about Safety**.

At this point you are probably thinking to yourself that the sinking ship scenario just discussed is a ridiculous example to use here especially in this discussion of laboratory safety related to laboratory design. You might be right! Hopefully, however, the reader does not confuse the example with the message. The message is clear, to the point, and has already been stated: Safety First and Early in Design.

To gain better appreciation for the message being delivered in this chapter, it would be wise to consider the following example which clearly demonstrates the absolute need to consider *safety first and early in the design process*.

10.2 LABORATORY DESIGN: AN EXAMPLE

Lex Stokes worked for an architectural firm that designed all types of business structures. Lex was recently assigned to a project to design a 100,000 square foot environmental laboratory. In reviewing the design specifications, Lex discovered that the only owner specifications provided included the requirement for 24 separate laboratory spaces to be built within the lab. Each separate lab space had certain specifications depending on the type of work to be performed in each space (e.g., metals lab was to be outfitted with equipment different from that installed in the microbiology lab). Another owner stipulation was that the laboratory must to be built to meet the latest safety and health regulatory compliance requirements.

Lex Stokes was an energetic, highly qualified young man who was determined to make his mark in the world of architecture. Thus, when his boss assigned him the environmental lab project, his first independent job assignment, he could not wait to begin his work.

Lex Stokes had a problem, however. Actually, he had two problems. In the first place, Lex Stokes had never worked on a laboratory design project; this would be his first one. In the second place, Lex Stokes was a loner who worked in isolation; he preferred to work in this manner.

The first problem, the fact that Lex Stokes had never designed an environmental laboratory, was not really that big of a deal—Lex Stokes was a college graduate—he had been trained to find information needed to finish any project; thus, he was equipped (educated) to overcome the lack of familiarity with environmental lab design.

In regards to information available on designing laboratories, several excellent publications and other standard references are available on the subject. For example, Lex could have consulted the excellent design specification text, *Guidelines for Laboratory Design* by Louis DiBerardinis et al (1993). In addition, texts dealing with building, life safety codes, and several basic design references are available. For example, the National Fire Protection Association (NFPA) Codes applicable to laboratory design are listed in Table 10.1.

TABLE 10.1. NFPA Codes Applicable to Laboratory Design.

NFPA 30	Flammable and combustible liquid code
NFPA 45	Standard on fire protection for laboratories
NFPA 70	National Electrical Code
NFPA 90A	Standard for the installation of HVAC systems
NFPA 91	Standard for the installation of blower and exhaust systems for dust, stock and vapor conveying or removal
NFPA 101	Life safety code—for safety from fire in buildings and structures

In addition to NFPA Codes the laboratory architect has the National Academy Press's *Prudent Practices Texts for Laboratories* (1981, 1983) available. Whenever construction of a building or structure is the goal, recognized and approved building codes such as *Building Officials and Code Administrators (BOCA)* should be consulted. At the very least the architect should consult Everett and Hughes' excellent and highly regarded work: *A Guide to Laboratory Design* (1981). Because Lex's client had specifically called for the new lab to meet all the current safety and health requirements, it would also be prudent for Lex to consult *Code of Federal Regulations*, Chapter 29, Part 1910, Most Recent Edition. This regulation lists OSHA's safety standards related to construction requirements dealing with:

- means of egress requirements
- walking and working surfaces
- ventilation requirements
- engineering controls to reduce noise
- machine guarding
- requirement for eye wash and safety showers
- restroom—minimum number of toilet facilities
- design of flammable and combustible liquid storage areas

Another source of safety related information or specifications for equipment installed in new construction (or anywhere else) are American National Standards Institute (ANSI) Standards. Table 10.2 contains a partial listing of ANSI Standards that should be consulted in the design of an environmental laboratory.

Lex Stokes had ready access to all of these references. This was not a problem. However, even though he used several of the above mentioned standards and codes in crafting his initial design, a problem did develop. Lex used these various references to design a 100,000 square laboratory with 24 separate cubicles (labs). He also included non-laboratory areas which included a conference room, eating/break area, equipment storage, gas cylinder storage library, maintenance room, offices, training room, and a copy room. The problem was that Lex Stokes forgot a standard rule of thumb. That is, he forgot that when designing something (anything) for human use it is impossible to redesign the human to use the machine, equipment, component or workplace. Instead, it is more practical to design the machine, equipment, component or workplace to fit the human (a basic concept in ergonomics).

TABLE 10.2. ANSI Standards Applicable to Laboratories.

ANSI/IES RP1	Practice for Office Lighting
ANSI/IES RP7	Practice for Industrial Lighting
ANSI Z136.1	Safe Use of Lasers
ANSI Z53.1	Safety Color Code for Marking Physical Hazards
ANSI Z87.1	Practice for Occupational and Educational Eye and Face Protection
ANSI Z358.1	Emergency Eyewash and Shower Equipment
ANSI/UL 924	Safety Standard for Emergency Lighting Equipment
ANSI A117.1	Specifications for Making Buildings Accessible to, and Usable by, Physically Handicapped People
ANSI A58.1	Minimum Design Loads for Buildings and Other Structures
ANSI/UL 399	Safety Standard for Drinking Water Coolers
ANSI/UL 199	Safety Standard for Automatic Sprinklers for Fire Protection Service
ANSI A14.3	Safety Requirements for Fixed Ladders

While it is true that Lex Stokes had all the correct references at his disposal to use (and he used a few) in the design of the environmental laboratory facility, it is also true that he had never been inside a laboratory or talked to anyone who had been inside an environmental laboratory. This, of course, was a colossal mistake.

Environmental laboratories are complex, include many technical areas, and are difficult to design. Designing an environmental laboratory in isolation like Lex Stokes did made the job exponentially more difficult. What Lex Stokes should have done was to consult the **user**. The user usually has a pretty good idea of what he/she expects in and from a particular work space. In addition to consulting the environmental laboratory user, Lex Stokes should also have consulted with the design engineer, human factor engineer, electrical engineer, industrial hygienist, toxicologist, legal representative, lab management (the user) and other architects.

Because Lex Stokes was not a seasoned laboratory designer and because he worked in isolation, several mistakes were apparent in his initial design work. The following list provides a few examples of the types of design mistakes Lex Stokes made:

(1) Ventilation—in general, fume hoods and vented work stations are technical areas which frequently present serious problems during the design of laboratories. Lex Stokes designed the ventilation exhaust system

for the environmental laboratory in such a way so as to be aesthetically pleasing to the eye as viewed from the outside. What this really means is that Lex Stokes installed the inlet (air intake) for fresh air in a location that could not be seen; it was also downwind of the fume stacks. Moreover, the 26 fume hood exhaust stacks were shortened to below roofline level (hidden) so they could not be seen from the street. The problem was his design caused fume recirculation problems.

(2) Because Lex Stokes new little about hazardous materials, emergency response, and even less about containment, he forgot to design door gullies in each doorway opening to lab areas where hazardous chemicals were to be used.

(3) Because Lex Stokes was not a safety specialist or industrial hygienist, he failed to plan (design) for required safety equipment. For example, he made no provision (available wall space) for the installation of portable fire extinguishers.

Lex Stokes, as a result of his design work on the environmental laboratory project, learned many valuable lessons. Probably one of the most important lessons he learned was that when one attempts to design anything, it is always appropriate and definitely prudent to seek professional expertise and guidance when it is required.

10.3 DESIGNING THE ENVIRONMENTAL LABORATORY— CORRECTLY

To avoid the mistakes architect Lex Stokes made in the preceding example and to design an environmental laboratory that will not only be user-friendly but also safe, it is probably wise to review and follow the advice provided by Bretherick (1991) in the following: "The three objectives in designing safe laboratories are (1) to provide a workplace which is suitable for the intended purpose(s); (2) to allow for human factors; and (3) to include necessary intrinsic safety features so far as is reasonably practicable" (p. 3).

After reviewing Bretherick's sound advice, the next move the design architect should take is to solicit the full cooperation of laboratory management, other architects, consulting engineers, safety professionals, and *laboratory users*. Emphasis is placed on the ***users*** here because it is for the users that the design should be tailored to fit. According to Schaefer (1991), "The user may not have any special expertise in laboratory design and construction, but he brings an essential ingredient to the process: he knows the laboratory has to work properly" (p. 9). The main point is that the design effort should be an integrated effort. Lex Stokes demonstrated that design conducted in isolation can bring about substandard results (this is a nice way of saying that he screwed up, bigtime).

When the architect who is tasked with designing the environmental laboratory knows what features to include (and what pitfalls to avoid) in his/her layout of the building, the corridors, and the rooms, he/she is well-suited to begin the actual design process.

In starting the environmental laboratory design process, the architect needs to solicit the help of the professionals mentioned earlier, enlist the advice and criticism of users and follow a protocol that will list the important elements to be considered in the design process. DiBerardins et al. (1993) suggest that this laboratory design protocol should include three general elements: (1) layout, (2) HVAC, and (3) occupational safety and health considerations.

The laboratory **layout** element should include several sub-elements. These sub-elements should include (1) personnel entry and exit provisions; (2) locations of laboratory furniture (e.g., benches, aisles, desks, and work surfaces); (3) location of fume hoods; (4) location of standard laboratory equipment; and (5) layout of handicapped access.

Along with chemical handling, storage, and waste disposal, HVAC, ventilation (in general), hoods and vented work stations are the technical areas that frequently present the most problems during the design of safe laboratories. These problems are the result of the complexity of the design and the delicacy of the internal environment that is to be maintained. This laboratory **heating, ventilation,** and **air conditioning (HVAC)** design element includes several sub-elements. These sub-elements include (1) designing temperature control equipment; (2) determining the proper laboratory pressure relationship (negative vs positive pressures); (3) design of laboratory ventilation systems that include comfort ventilation supply for laboratory modules and recirculation of laboratory room air; (4) design of exhaust ventilation for each laboratory module, for general room ventilation air from laboratories, for proper air rates to hoods and other local exhaust air requirements; and (5) design of proper-sized exhaust fans and blowers, air cleaning equipment for laboratory effluent air, and for exhaust ducts and plenums.

In the design element that covers the **occupational safety and health considerations**, several factors must be addressed. For example, factors such as emergency considerations for ground fault circuit interrupters, emergency fuel gas shutoff, master electrical disconnect switches, emergency blowers, chemical spill control, and emergency eyewash stations all need to be designed into the laboratory facility. In addition, safety and health considerations necessitate ensuring that proper construction materials and methods are employed, various control systems are designed to specifications, and alarm systems for laboratory equipment are part of the design process. It is also important to make design considerations that

TABLE 10.3. Design Criteria for Environmental Laboratories.

Criterion	Explanation
1. Creation of space that is stimulating	Design should include natural lighting and open views to the exterior to preclude fatigue and thus enhance safety (see Figure 10.1)
2. Worker protection	The design must minimize the risk of worker exposures.
3. Traffic flow	The design must consider the patterns of movement of both people and material.
4. Fire protection	The design should include a total coverage sprinkler system for the entire facility. Each lab space should have rate-of-rise heat detectors. Fire alarm pull boxes must be included for location in and along the laboratory corridor.
5. Air supply and exhaust	The design should include a dedicated air supply system that can provide temperature and humidity control throughout the year. The design should also include an exhaust system that will provide at least 10 air changes per hour (ach). In addition, the system must maintain pressure differentials relative to ambient air as per specifications.
6. Lighting	Fluorescent lamp lighting is recommended for laboratory spaces.
7. Maintainability	Critical mechanical systems must remain functional; thus, it is necessary to design sensing and alarming devices to indicate system failure.
8. Storage	The lab design must include space for chemical and chemical waste storage.
9. Human factors	The lab design should include using the principles of human factors engineering to provide a comfortable work environment that minimizes stress.
10. Cleanable	Epoxy finishes should be used on surfaces such as walls, floors, and ceilings to facilitate easy cleanup of spills; moreover, these finishes need to be resistant to chemical attack.

will address safe storage of hazardous materials, compressed-gas cylinders, and installation of required safety equipment.

Along with consideration of these important elements of design related to laboratory layout, HVAC, and occupational safety and health concerns, the project architect, based on the design review and user/needs assessment, should formulate a basic design criteria for the project. Again, an excellent reference for putting together this design criteria is ***Guidelines for Laboratory Design*** by Louis DiBerardinis et al. (1993). A criteria that will include the necessary criterions to design the environmental laboratory can be developed from this source, some elements of which are shown and explained in Table 10.3

Figure 10.1. Environmental laboratory fume hood exhaust stacks. Some might say these stacks present an eyesore; others, (industrial hygienists and designers) would say that someone used good engineering/design practice here.

CHAPTER 11

Ventilation

Air is lazy; it goes where directed and rarely thinks for itself. (Saunders, 1993, p. xiii)

11.1 INTRODUCTION

It is because air is lazy, goes where directed, and rarely thinks for itself that a dedicated, highly sophisticated ventilation system is required for the laboratory. This is especially the case in laboratories where control of chemical exposure to lab technicians is required.

As important as laboratory ventilation is it is surprising how little attention is given to the topic and how little is understood about it. This is generally the case because lab workers are not trained on the lab's ventilation system and its controls. The intent of this chapter is to correct this educational deficiency by providing a basic review of typical laboratory ventilation systems and pointing out the importance of informing all lab workers about their lab's ventilation system and its controls.

11.2 VENTILATION BASICS

In order to gain a fundamental understanding of ventilation it is necessary to investigate the basics. Probably the simplest way to describe *ventilation* is to simply say that it is *the designed and purposeful movement of air*. Obviously, ventilation begins with *airflow*. Air, defined in its simplest terms or form, is an incompressible fluid that is invisible and non-tactile; because of these qualities it is often ignored, disregarded. When attempting to visualize airflow, it is helpful to refer to the other incompressible fluid, water. By making airflow analogous to fluid flow, it is easier to comprehend the significance of turbulence caused by obstructions or obstacles within the fluid flow. This turbulence is not insignificant.

To help explain the significance of turbulence and its impact on airflow and on those around or near the airflow, Knutson (1991) uses the example of the rock in a fast moving stream. In the stream the rock forces the water to flow around it, which forms a highly turbulent area downstream of the rock. Vortices are formed causing the water to move upstream toward the rock.

Knutson (1991) goes on to make the analogy clearer by pointing out that in a similar manner as the action caused by the rock in the stream, air flowing past a person will cause a negative depression downstream which will cause the formation of backflow eddies and vortices. The importance of this can be seen as the air moves past a person in front of a laboratory fume hood or other exhaust ventilation system. Obviously, when the vortices form near the work and entrain contaminants generated within the hood and the backflow eddys draw these contaminants toward the lab technician at the hood face, significant exposures can result. This type of result certainly would be ironic when you consider that ventilation is designed to reduce employee exposure to airborne contaminants and not to increase their exposure.

11.3 VENTILATION SYSTEMS

A typical environmental laboratory has several types of ventilation. Exhaust ventilation systems are of two types: (1) *general* exhaust systems and (2) *local* exhaust systems. Ventilation can be used either to capture contaminants and remove them or to dilute contaminants to safe levels.

11.3.1 GENERAL EXHAUST (DILUTION) VENTILATION

General exhaust or dilution ventilation can be used for both heat control and/or contamination removal. Dilution occurs when contaminants mix with air flowing through the work area. Usually, contaminants are diluted by mechanically induced air movement with a large volume of outside air drawn in and mixed with it. The contaminated air is then discharged to the outside atmosphere. The major disadvantage of dilution ventilation is that a large volume of dilution air may be needed. Thus, this system is normally used for contaminant control only when local exhaust is impractical.

11.3.2 LOCAL EXHAUST VENTILATION

In general exhaust ventilation the objective is to dilute the contaminant; on the other hand, in local exhaust ventilation the objective is to *capture* the contaminant. Local exhaust ventilation is the preferred method of control because it is more cost effective relative to dilution ventilation, where huge volumes of air can be moved at any given moment. In addition, local exhaust ventilation is usually more economical whenever it is tied into the facility's HVAC system where return air must be preheated. Obviously, it is less costly to preheat small volumes of return air as compared to huge volumes that are typical with general exhaust ventilation systems. Another huge advantage of local exhaust ventilation, especially in the laboratory environment, is that it provides better control of the contaminant. This greater control is obtained by having the exhaust device surround or enclose the source; as would be the case in a laboratory fume hood. Of course, the benefit received from having greater control of the contaminant is that gases and particles that are small enough to be breathable normally follow the airflow, control of the airflow provides control of the contaminant and protects the user. Simply stated, local exhaust systems work to capture or contain contaminants at their source before they escape into the workplace environment; thus, they work to protect (by providing positive control over exposure) those who work in the laboratory.

Local exhaust systems consist of such devices as bench slots, canopy hoods, elephant trunks, backdraft benches, and laboratory fume hoods. It is the laboratory fume hood that is the focus of discussion in this chapter.

A typical local exhaust system consists of five components: hoods, duct system, air cleaning device, fan and an exhaust stack.

A *hood* is a shaped inlet designed to capture contaminated air and conduct it into the exhaust duct system; the hood is the most important part of the local exhaust ventilation system (see Figure 11.1). The purpose of the hood is to collect the contaminant generated in an air stream directed toward the hood. Both the design and location of the hoods are crucial in determining whether the system will work as expected. There are three types of hoods, each working on a different principle:

- **enclosure hood**
- **receiving hood**
- **capturing hood**

The **enclosure hood** surrounds the contaminant source as much as possible. This type of hood has a low exhaust rate with an inward face velocity of between 100–150 ft/min. Contaminants are trapped inside the enclosure by air flowing in through openings in the enclosure.

The **receiving hood** is used for processes that generate a waste stream of contaminants in one specific direction. For example, a blast furnace that emits a hot stream of air and gases above it would need an exhaust system mounted directly above it. This would facilitate catching the contaminants when they are generated. These hoods are generally not used where health protection is an issue. This is the case because these hoods operate on low velocity airflow. When the airflow is low velocity gases, vapors, and very small particles that do not travel very far, high velocity moving air is required.

The **capturing hood** actually reaches out into the workplace to capture contaminants in the air. These hoods are widely used in industry because they can be placed alongside the contaminant source rather than surrounding it with an enclosure hood. Capturing velocities as low as 50 ft/min to as high as 2,000 ft/min are typically used. The capturing hood does have two major disadvantages: (1) Large air volumes may be needed to generate an adequate capture velocity at the contaminant source; and (2) the reach of most capturing hoods is limited to about 2 feet from the hood opening. This second disadvantage is in line with the well-known ventilation adage that says: It is easier to provide a stronger blowing force (supply) than it is to provide a strong sucking force (exhaust).

In selecting the proper hood for the environmental laboratory it is important to keep a couple of factors in mind. First, each hood type works on a different principle and has a very distinct purpose. Second, selecting the proper hood will depend on the type of contaminants that are or will be generated inside the environmental laboratory. Gases, vapors, and particulate matter in respirable size go with the flow—that is, they move along with the airflow. Contaminants are thus controlled by regulating the airflow around them.

Ducts are a system or network of piping that connects the hoods and other system components. An *air cleaner* is a device that removes atmospheric airborne impurities such as dusts, gases, vapors, smoke, and fumes. The *fan* provides the motive force (energy) to draw air and contaminants into the exhaust system by inducing a negative pressure or suction in the ducts leading to the hoods. The *exhaust stack* provides a safe outlet for contaminants to the atmosphere. These stacks are normally positioned in such a manner to preclude exhausted air and contaminants

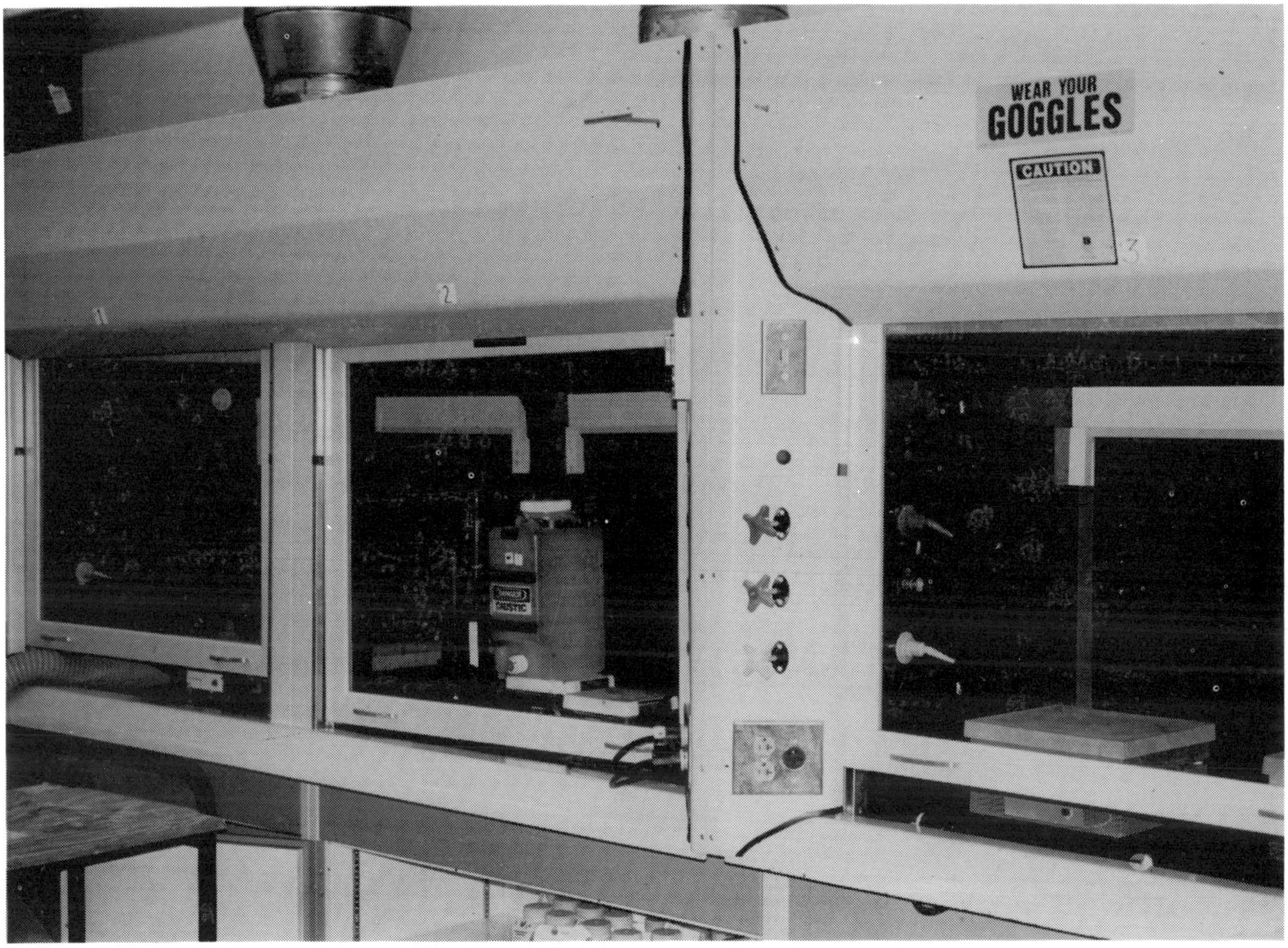

Figure 11.1. A series of fume hoods of the type used in environmental laboratories.

being re-circulated with outside supply air (see Figure 11.2).

11.3.3 HEAT CONTROL VENTILATION

Heat control ventilation (sometimes called comfort ventilation) is used to control indoor atmospheric conditions associated with hot industrial environments such as are found in laundries, foundries, bakeries, etc., for the purpose of preventing worker discomfort. In many laboratories, there are significant heat sources which can cause an elevation of the workplace temperature. Obviously, this is not a good or comfortable situation for the workers. Thus, exhaust ventilation is provided to control the heat and humidity to provide comfort to lab technicians.

11.3.4 SUPPLY VENTILATION

Whenever several devices such as fume hoods work to exhaust air from the laboratory, obviously an equal quantity of air must be supplied. If the air supply system is inadequate (under-designed or inoperative) then supply air will get inside the laboratory through infiltration, backdrafting in inactive hoods, or by some other undesirable method. Consequently, a dedicated air supply system is required to ensure proper operation of the exhaust ventilation system and workplace comfort.

11.3.5 THE IDEAL MIX

Even though supply or return air is required to provide replacement of air that has been exhausted, the supply and exhaust systems in the typical laboratory work on two different principles. Gentle removal of contaminants is the function of exhaust systems. Supply systems, on the other hand, provide control of environmental conditions by thoroughly mixing the air in the room. Because supply causes churning or mixing of the air and because exhaust systems have to overcome these actions to properly remove contaminants, there must be an ideal mix or balance between the supply and exhaust ventilation system. Under normal conditions, it is desirable to maintain a balance between supply and exhaust—to have approximately the same supply and exhaust for each laboratory unit. However, there are other instances where it

(a)

(b)

Figure 11.2. Different views of fume hood stacks. These stacks are high enough and positioned in such a way to preclude exhaust being taken up by lab intake ports.

(c) 

Figure 11.2. *Continued*

may be desirable to provide an intentional imbalance. For example, there may be cause to provide positive or negative conditions in the laboratory to provide a secondary level of control. For instance, when more air is exhausted from a laboratory than is provided through the supply air system, the room is "negative." When this occurs, air will flow through openings and there will be a migration from the corridors into the laboratories. This might be the desired effect. That is, under a significant chemical or contaminant release or spill condition, the negative condition will maintain the chemical or contaminant within the laboratory without exposing outside personnel.

11.4 SAFE WORK PRACTICES: USING LABORATORY FUME HOODS

Before beginning a detailed discussion of recommended safe work practices to be observed and performed with fume hoods in the environmental laboratory, it is important to discuss a few basic fundamental concepts. For example, at the heart of safe operation of fume hoods is the ***face velocity*** of the hood. **Face velocity** refers to the average air velocity into the exhaust system measured at the opening into the fume hood.

OSHA, under its 29 CFR 1910.1450 (Laboratory Standard), is very specific about maintaining the face velocity of laboratory fume hoods within certain ranges, depending on the toxicity of contaminant. The higher the toxicity (Threshold Limit Value—TLV) the higher the face velocity required to ensure adequate protection for the worker. At the minimum, OSHA requires laboratory fume hood face velocities of 60 to 100 ft/min. A list of the required face velocity levels for use with chemicals with varying levels of toxicity in fume hoods is provided in Table 11.1.

TABLE 11.1. Face Velocity for Lab Fume Hoods.

Chemical Toxicity Level	Required Face Velocity Range
Chemicals of low toxicity	70–80 fpm
Chemicals of moderate toxicity	100 fpm
Chemicals of high toxicity	125–150 fpm

When working with laboratory hoods, workers—in order to receive the full protection afforded by the hood, should follow the safe work practices listed in the following:

(1) Chemical operations that release significant quantities of airborne contaminants should be conducted inside the hood.
(2) The user should keep all containers and apparatus several inches back from the front of the hood. Moving an apparatusseveral inches back from the front

edge of the hood can reduce the vapor concentration at the face by 90% (Hall, 1994). *Note:* It is important to keep in mind that the performance of the hood depends on placement of equipment in the hood; room drafts from open doors or windows; people walking by; and even the presence of the user in the front of the hood.

(3) Keep head outside hood enclosure when operations are underway.
(4) Except for quantities of volatile materials of less than 50 ml, the fume hood should not be used for evaporating waste fordisposal purposes.
(5) Do not store chemicals in hoods; chemicals should be stored in approved storage cabinets.
(6) Keep the hood sash closed as much as possible. Keeping the face opening of the hood small will improve the overall performance of the hood.
(7) Do not obstruct the baffles or slots in the rear of the hood.
(8) If there is the possibility of an explosion, a safety shield or barricade must be used.
(9) Keep laboratory windows and doors closed.
(10) Control traffic past the hood.
(11) When in use, the hood must be "on."
(12) Except for repair or routine maintenance, do not remove any component part of the hood.
(13) Only use a fume hood that is approved for use (i.e., one that has been properly maintained and tested).

11.5 TESTING HOODS

Periodic checks or tests of laboratory fume hood performance are important in protecting workers' health. With normal use the hood's efficiency in containing or capturing contaminants may decrease for many reasons: normal wear; loose component parts (e.g., fan belts); out-of-adjustment dampers; clogged air cleaners; and duct-work blocked by dust and debris.

It is important to keep in mind that even though the fume hood equipment might be performing up to standards it may still not provide enough protection for the user. The point is that no matter how well the unit checks or tests out, the system is not performing adequately if air samples show excessive employee exposures. One can only be sure that the hood system is providing sufficient protection if hood system tests can be correlated with air contaminant concentrations.

Several different tests are available to help determine how the hood system is operating. These tests consist of: hood static pressure; Pitot tube duct velocity measurements; velocity at the hood opening; smoke tube tracers; and fan pressure, rotating speed, and power consumption. Along with performing such tests on the hood system to check how it is operating, routine maintenance should also be performed on a regularly scheduled basis.

11.5.1 DAILY AND BEFORE USE INSPECTIONS/ TESTS

At a minimum daily/before use and quarterly inspection and testing should be conducted on laboratory fume hoods. The daily, or before operation inspection, should consist of an *airflow check.* A simple qualitative check that can be used to check for proper airflow, exhaust, is the ribbon or tissue paper check. During this test, a ribbon or piece of tissue paper (be careful not to lose control of these for they could "gum-up" the system) is placed at the hood opening to determine if it reflects directional airflow. Daily and before each use the hood gages and monitors should be checked for proper operation within a predetermined range of operation.

Daily and before each use the hood should be checked to ensure that material is not blocking the hood opening or exhaust ports. This housekeeping procedure should also include checking the fume hood to ensure that there are no puddles of spilled materials pooled within.

11.5.2 QUARTERLY INSPECTION AND TESTING

Face velocities for laboratory fume hoods should be measured on a quarterly basis. These tests should be conducted by qualified personnel using a properly calibrated velometer.

Smoke tube or smoke candle tests should be performed quarterly to evaluate airflow patterns.

A final note. The practice of conducting these daily and quarterly inspections and tests on laboratory hoods should be made part of the laboratory's safety training program. In addition, to ensure maximum protection, workers should observe the safe work practices outlined above.

PART V

PROTECTIVE EQUIPMENT

CHAPTER 12

Personal Protective Equipment (PPE)

Of the nearly two million people expected to receive disabling work-related injuries this year, more than 25 percent of these injuries will involve the head, eyes, hands, or feet (Spellman, 1996, p. 69)

12.1 INTRODUCTION

On October 5, 1994, the Occupational Safety and Health Administration (OSHA) issued regulations governing the use of personal protective equipment (PPE). These regulations, which were recently updated to include the latest American National Standards Institute (ANSI) criteria for effective workplace protection, are found at 29 CFR 1910.132-.138, 1910.1000, 1910.1001-1045, and 1910.95 (Hearing Conservation). Although laboratory managers should keep in mind that PPE should not be used as a substitute for engineering, work practice, and/or administrative controls, these regulations directly impact the worker and work environment in the environmental laboratory (Spellman, 1996). This is the case, of course, because laboratory workers face the constant risk of exposure to a wide variety of hazardous substances.

The PPE addressed in the OSHA Standard specifically deals with protective equipment designed to protect many parts of workers' bodies including eyes, face, head, hands, and feet. PPE for Respiratory Protection and Hearing Conservation are addressed in other OSHA standards.

In general, OSHA's PPE Standard mandates three requirements:

- The laboratory Safety Officer or Organizational Safety Director must conduct both a work center and a job classification *hazard assessment* to determine (1) where PPE is required, (2) which job actions require PPE, and (3) who is required to use PPE.
- Laboratory Management *must provide approved PPE* to personnel who are required to use it in the normal performance of their duties.
- The employer must *train* workers on where, how, and when to use PPE. In addition, workers must be trained on the limitations of PPE.

12.1.1 PPE: HAZARD ASSESSMENT

The hazard assessment requirement of OSHA's Personal Protective Equipment (PPE) Standard basically affords the employer the opportunity to choose the PPE that is appropriate for each particular workplace situation. In addition, it also affords the employer the opportunity to look at certain work practices, in the past, that have shown the tendency to cause a significant number of accidents and injuries; this enables the employer to make changes in these outdated, inefficient, and unsafe work practices.

The first step in determining the types of PPE that are needed in the environmental laboratory is to conduct a hazard assessment survey to determine if hazards that require the use of PPE are present or are likely to be present. Along with conducting this walk-around survey, workplace injury and accident data should be reviewed. Experience has shown that a walk-around survey such as the one needed to conduct a viable hazard assessment survey will be more productive if the workers are informed of what is going on. Informed employees can be a valuable resource. When you get right down to it it is the employees who will have to wear the PPE; thus, it is they who will benefit from its use. Moreover, since they may have to use PPE, it only makes good sense to bring them into the decision-making process.

After making the walk-around survey of the laboratory and writing down the findings, the collected data should be analyzed. It should be pointed out here, as was mentioned briefly earlier, it is important not only to look at the physical plant and equipment within the environmental laboratory but also at the **work** the lab technicians perform on a daily basis. These daily work practices need to be looked at closely to determine not only what PPE is called for but also to determine if the work practice is a safe procedure.

Once the data from the walk-around survey is analyzed, then the current PPE used should be reviewed. In work practices where additional PPE is required it should be selected and issued to the workers. It is important to point out that the PPE issued to workers must *fit* the worker. Giving workers the wrong size protective gloves, safety shoes, eye protection and other PPE is not allowed under the PPE Standard.

The PPE hazard assessment procedure is an on-going process. The initial hazard assessment and walk-around establish the baseline signature of the workplace and work practices conducted in the workplace. Excepting the premise that all things change with time, it stands to reason that the laboratory environment and work practices conducted within will change with time. Changing environments and work practices might call for changes in PPE requirements. Thus, the PPE hazard assessment procedure should be ongoing (conducted on an annual basis).

The PPE hazard assessment must be verified through written certification. Figure 12.1 shows a sample PPE Hazard Assessment Form. This form not only documents hazard assessment data but also provides written verification that the assessment was conducted.

12.1.2 THE EMPLOYER MUST PROVIDE APPROVED PPE

Under OSHA's PPE Standard, the employer must not only determine the type of PPE required for the types of work performed in the workplace it must also select approved PPE for workers' use. As a case in point, for example, let's take a look at what is required of the employer in regards to providing workers with proper PPE in the environmental laboratory. First of all, lab workers should have ***eye protection*** available that consists of safety glasses, goggles, and face shields. It should be pointed out that when performing work with liquid chemicals the lab worker should wear both goggles and a face shield. It should also be pointed out that a face shield *is not* eye protection; instead, it is face protection.

OSHA requires that eye protection meet the following minimum requirements:

- provide adequate protection against the particular hazard for which they are designed
- be reasonably comfortable when worn under the designated conditions
- fit snugly without interfering with the wearer's movements or vision
- be durable
- be capable of being disinfected
- be easily cleanable
- be kept clean and in good repair

Safety glasses are the basic form of eye protection (see Figure 12.2). OSHA now requires protective eye coverage from the front and the sides any time there is a hazard from flying objects. Such coverage can be accomplished by using safety glasses with attached side shields or by using detachable side protectors.

Many workers wear prescriptive eyewear on the job. When these workers are required to wear eye protection in the lab, they must also wear approved eye protection. Some workers who wear prescriptive glasses wrongly assume that they are automatically wearing eye protection. This is not the case, however. It should be pointed out that even though OSHA requires employers to provide eye protection for workers it does not require employers to purchase or pay for workers' prescriptive type safety eyewear (Often this is a point of confusion and sometimes contention for many workers; thus, it should be clearly explained to the worker during the training session on PPE). If the worker desires to purchase his/her own protective eyewear ground to his/her prescription, the protective eyewear must meet the applicable ANSI (American National Standards Institute) standard. For laboratory work, it is wise to issue workers with goggles that are oversized; that is, designed to fit over their regular prescription glasses. This action seems to suit workers' needs while also providing the required protection (Spellman, 1996).

Contact lenses should not be worn in the environmental laboratory. Vapors and gases can be concentrated under such lenses and cause permanent eye damage. Further, in the event of chemical spill where chemical enters the eye, it is nearly impossible to remove a contact lens quick enough to properly irrigate the eye. Another factor to keep in mind (and definitely out of your eye) is that solvent vapors readily absorb into soft contact lenses. This situation, of course, is not good.

Laboratory workers who work with bottles of compressed gases, for example, and/or are required to lift and carry large trays of glassware or chemical containers from place to place should be required to wear safety shoes. It is interesting to note that the employer is required by OSHA to ensure that the worker wears safety shoes when they are required but is not required to purchase the shoes for the worker. Again, this can be a point of confusion and contention for workers—PPE training should clarify their requirement and the employer's responsibility regarding safety shoes.

PPE HAZARD ASSESSMENT FORM

1. Job Classification:
2. Location of job, function, activity or situation:
3. List all apparent hazards
    a. Impact
    b. Penetration
    c. Compression (roll-over)
    d. Chemical
    e. Heat
    f. Harmful Dust
    g. Light
    h. Noise Levels
    i. Respiratory
4. Personal Protective Equipment now provided for this work practice:
5. Additional PPE required for safe work practice:
6. PPE Training:
7. PPE Maintenance function, specifications, duties:

____

Assessment Completed by: (Print Name)

____

Signature Date

____

Title

Figure 12.1. PPE hazard assessment form.

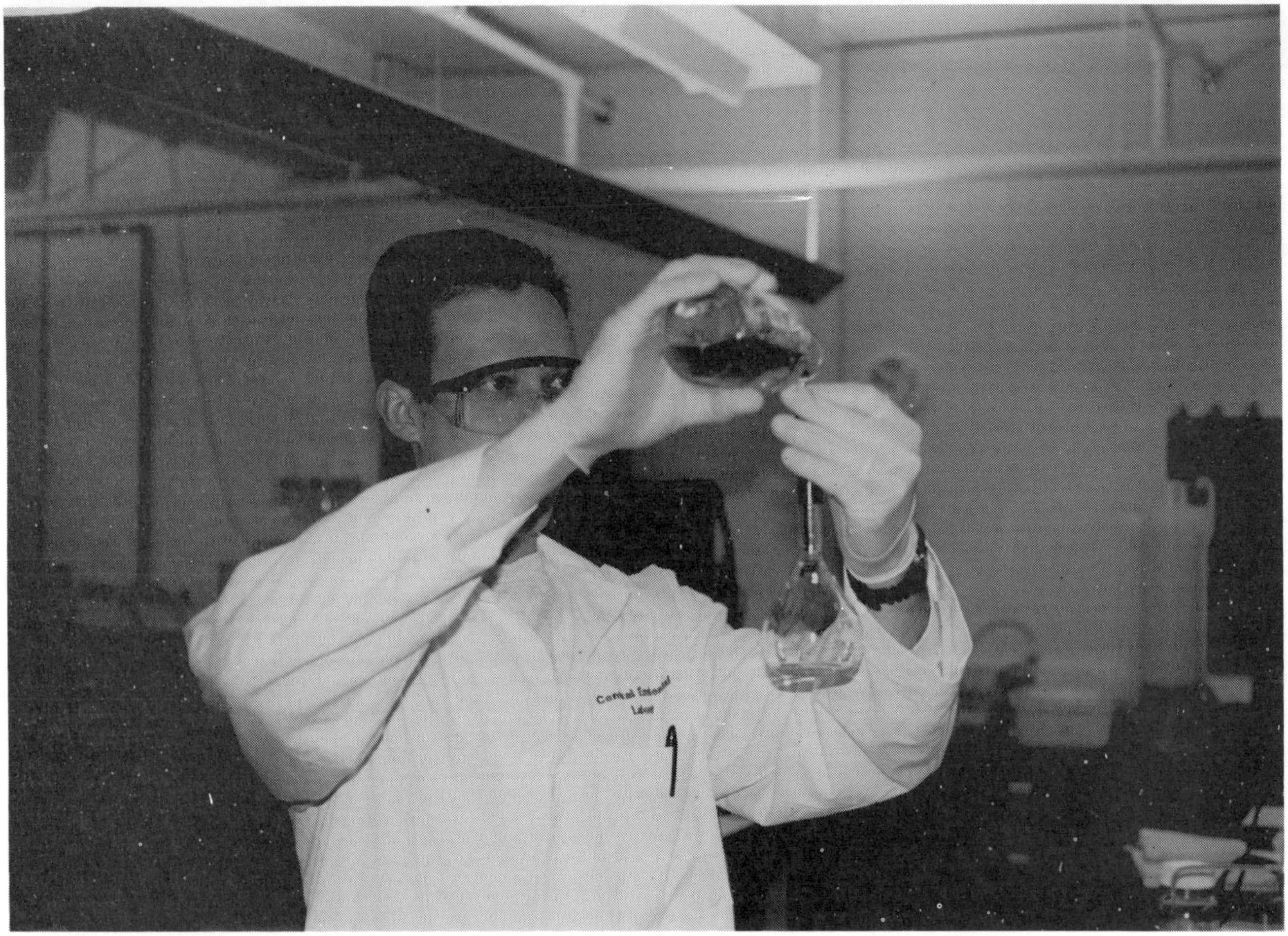

Figure 12.2. Lab specialist wearing proper personal protective equipment; notice that the safety glases include side shields.

Next to the universal and absolutely essential need to wear eye protection in the laboratory, the need to wear some type of ***hand and arm*** protection is just as important. Hand and arm protection devices that are required to protect workers on the job must be provided by the employer. The key ingredient to proper hand and arm protection is to ensure that workers wear protection when required. In the environmental laboratory there are few work practices that do not require both eye and hand/arm protection. Again, exactly where and when this PPE is to be worn by workers will be determined during the hazard assessment.

In the laboratory environment the need for hard hat (head) protection is usually limited. However, if there is a need (determined through the hazard assessment), the employer must provide hard hats for workers' use.

If hearing and/or respiratory protection is required for use in the environmental laboratory, then the employer is responsible to provide this protection. Both Hearing Conservation and Respiratory Protection are covered in separate OSHA standards; they are not part of the PPE Standard.

12.1.3 PPE TRAINING

Training is at the heart of worker safety (see Figure 12.3). This point seems clear—valid—and appropriate when one considers that workers need to know what is expected of them, how to perform a particular function, and the hazards involved with each work practice to perform their jobs safely. Under OSHA's PPE Standard there is a definite requirement for worker training. This requirement makes sense when you consider that according the the Bureau of Labor Statistics' injury report, a significant number of workers who are injured on the job were never trained in the proper use of PPE.

What exactly should this PPE training consist of?

Good question!

This PPE training must inform workers on when to wear PPE; workers must know what PPE to wear; they must know how to wear the PPE; they must know the limitations of PPE; and, they must be trained in the proper care, maintenance, useful life and disposal of PPE.

It is important to point out that not only is PPE training required it must also be documented. Keep in mind that

Figure 12.3. Training is at the heart of safety.

training conducted without proper written documentation is training that was never conducted, in the eyes of the Regulator (Spellman, 1996).

In light of this training verification requirement, it is wise to create some type of certificate of training that can be awarded to the worker. This training certificate can also be copied and made a permanent part of the worker's training record (see Figure 12.4).

In addition to using a certificate of training to verify that the training was conducted it is also a good idea to have each worker sign and date an attendance roster. Figure 12.5 shows one version of such a roster.

A final point about PPE and employee training—an effective laboratory personal protective equipment program cannot exist without a mandatory and comprehensive employee training program. During these mandatory and comprehensive employee training sessions the employer must make it perfectly clear to all workers that not only does OSHA mandate the use of PPE in the workplace but so does the employer!

12.2 PPE AND THE ENVIRONMENTAL LAB

The PPE requirements discussed to this point relate to all work centers in general industry (the construction industry also has PPE requirements but it is covered under 29 CFR 1926). Since the environmental laboratory is usually classified as a general industry-type workplace, the PPE requirements covered to this point are fully applicable to the lab. An additional point should be made, however. The environmental laboratory will expose workers to substances, chemicals, and hazardous materials that are different from those that might be encountered in other general industry-type workplaces. Thus, additional requirements related to PPE are required in the laboratory-type workplace.

One of these additional or unusual (site-specific) requirements can best be explained, for example, by comparing safety goggles. Safety goggles used to protect the worker from flying debris from a grinding wheel or some other such device are usually vented so that the lenses

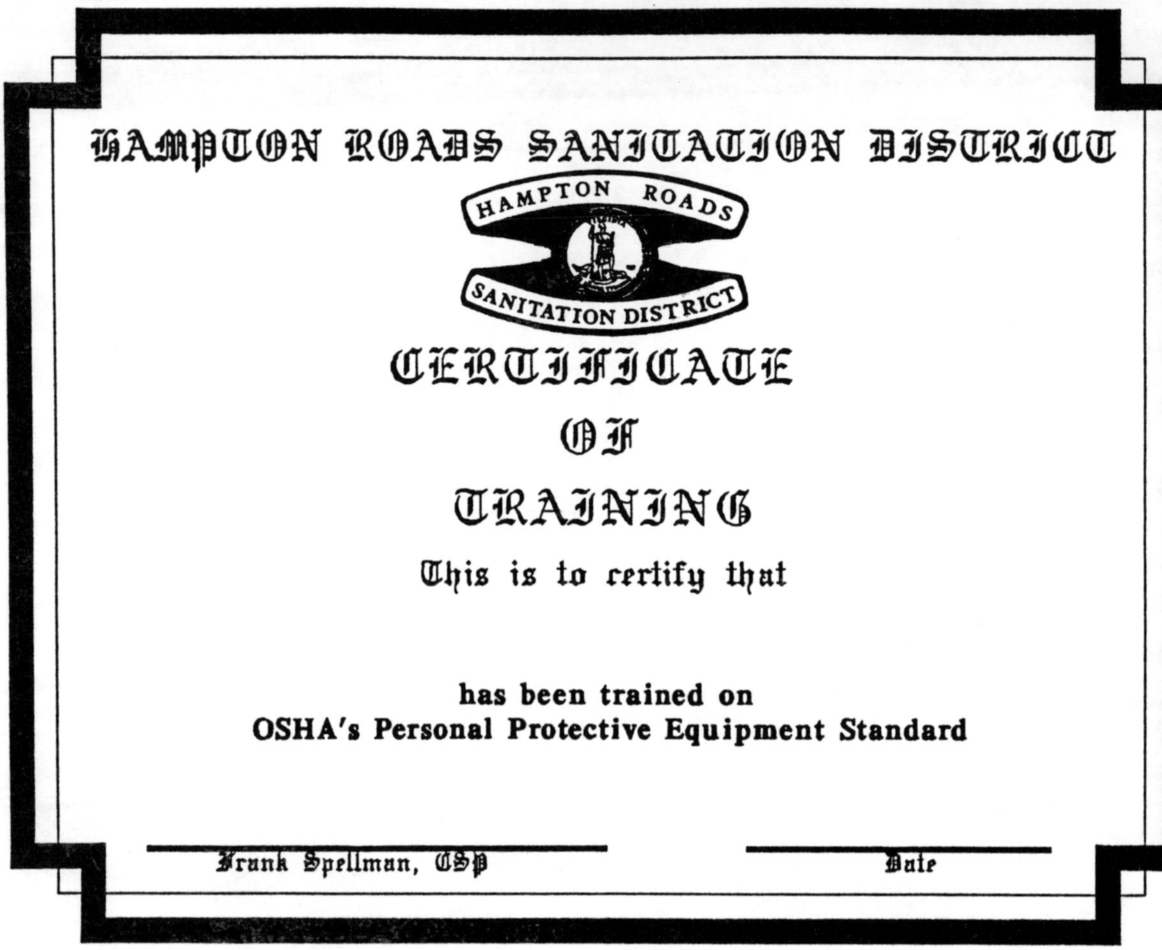
HAMPTON ROADS SANITATION DISTRICT

HAMPTON ROADS SANITATION DISTRICT

CERTIFICATE

OF

TRAINING

This is to certify that

has been trained on
OSHA's Personal Protective Equipment Standard

Frank Spellman, CSP

Date

Figure 12.4. A training certificate used to verify training was conducted. Courtesy of HRSD.

do not fog over. By installing such vents in standard safety goggles no problem is presented to the worker who is grinding—the same cannot be said, however, for the lab technician who is working with a hazardous gaseous substance. It would do little good, obviously, for the lab technician to use vented goggles when working with hazardous gases.

Another requirement (in regards to workers wearing PPE in the workplace) that the employer must take great care to ensure is met is the medical evaluation requirement. Medical evaluation of personnel for fitness for duty and for use of personnel protective equipment (such respirators) must be one of the criteria for selection and continuing certification of its use. More will be said about this important requirement, with regards to respirators, in Chapter 14.

12.2.1 BODY PROTECTION

Except for special tasks such as brazing & welding, for example, special clothing for full body protection is not normally required for workers in general industry-type workplaces. However, this is not the case in the environmental laboratory. Laboratory workers must pay particular attention to the clothing they wear in the laboratory because it can be important to their safety. Loose, torn, skimpy clothing should never be worn in the laboratory. Loose of torn clothing mixed with a close encounter with a bunsen burner creates a fire hazard. Moreover, loose and torn clothing can become ensnared in moving parts of laboratory equipment. In the event of chemical splash, the last thing any rational lab worker would want is excessive bare skin that is readily accessible for the chemical to reside on, to corrode, and/or to destroy.

A generally accepted rule that exists in most laboratories, related to clothing in the lab, states: if the lab worker has the potential to be exposed to chemicals, personal clothing that is to be worn home will be covered by protective apparel such as lab coats and aprons. In chemical emergency response situations the correct level of clothing as directed by OSHA guidelines should be worn. More will be said about protective clothing to be used in chemical emergencies later in this text.

ATTENDANCE ROSTER
SUBJECT MATTER: PPE

____

In accordance with the recordkeeping and training requirements of the Personal Protective Equipment (PPE) Standard, OSHA 29 CFR 1910. 132 and 1910.138, I have received training on when PPE is required, what PPE I must wear, how to wear the PPE, the limitations of PPE, and the proper care, maintenance, useful life and disposal of PPE. I have agreed to verify my understanding and training of 1910.132 and 1019.138 by signing and dating this form.

WORKER'S NAME	DATE
________________	________
________________	________
________________	________
________________	________
________________	________
________________	________
________________	________
________________	________

Figure 12.5. Training attendance roster used to verify the worker's attendance at a particular training session; it also serves as verification that the worker understood the training. This form or a similar form should be used for all saftey training sessions and should be filed in the workplace master training file.

CHAPTER 13

Safety Showers/Eyewash Stations

Safety showers and eye wash fountains must be at easily accessible locations for all employees working in the laboratory. The location should be at such a place that employees would not be tempted to place boxes or other items in such a position to restrict access to the shower or eye wash. (Phifer, 1991, p. 13)

13.1 INTRODUCTION

IN the previous chapter information was presented concerning the various types of personal protective equipment (PPE) required to protect laboratory workers. It was also pointed out earlier that putting complete reliance on PPE to protect workers is not recommended. Instead, engineering controls should always be incorporated first—the idea is to "engineer out the hazard," if possible.

Those who have worked out there in the "real world" know from experience that it is impossible to engineer out every safety and health hazard. With this point in mind, designers, architects, engineers, and safety professionals try to identify the hazards that cannot be engineered out and provide for reasonable accommodation. This accommodation comes about in providing suitable safety and emergency equipment—which should be strategically placed, for easy worker use, throughout the laboratory.

This chapter deals with two vital safety and emergency systems which are required to be installed in laboratories that use or handle chemicals: eyewash fountains and safety showers. Before beginning discussion of these two vital safety systems it should be pointed out that the standard environmental laboratory requires outfitting with several different types of safety equipment. In previous chapters and in those to follow most of this important safety equipment will be covered including standard laboratory safety equipment such as chemical spill kits and fire extinguishers. For now it should be emphasized that it is the responsibility of the Chemical Hygiene Officer (CHO) to recommend and provide the supplementary safety and emergency equipment as needed (Hall, 1994). In addition, the CHO is responsible to ensure that all safety equipment is operational at all times (remember Fred Kruger). This includes ensuring that all laboratory safety equipment is tested on a routine basis—and that it is repaired or replaced as needed.

13.2 EYEWASH FOUNTAINS

According to OSHA's 29 CFR 1910.151(c) of the General Industry Standards: *"Where the eyes . . . may be exposed to injurious corrosive materials, suitable facilities for quick drenching and flushing of the eyes . . . shall be provided within the work area for immediate emergency use."*

Whenever an employee receives an injury while on the job, this is a tragic event—an event usually caused when someone or something has failed in one way or another. For now this discussion is not concerned with causal factors but with preventive measures that can and should be employed in the environmental laboratory to prevent or reduce the possibility of injury.

During chemical safety training sessions that I have conducted over the years I always begin such a session with an "eye-opener", so to speak. I have found that when it comes to eye protection safety training, the trainer need only ask his/her audience a few simple questions: (1) How devastating would it be for you to lose your eyesight? (2) Other than loss of life, is there any other injury that is more devastating to you than loss of your eyesight? (3) When you talk to someone face to face and they look you in your disfigured eyes, how will this affect you?

These questions usually set the tone for the lesson that I subsequently present on the importance of eye protection. Experience has shown that laboratory workers must learn the location of and the use of available safety and emergency equipment in their work area so that he/she can find them with eyes closed.

With eyes closed?

Yes!

Absolutely.

This statement seems appropriate when you consider that when an employee fills his or her eyes with 30% sodium hydroxide solution, for example, his or her eyes will definitely be closed—if open he or she will be blinded by the corrosive chemical, by the probing irritation, and by the excruciating pain. The point is that the lab worker who is injured in such a manner will literally be blinded (eyes closed—vision impaired) and will have to find an eyewash station or fountain just as a blind person would.

The reader is probably thinking to him/herself: How could such a stupid incident occur in the first place—aren't lab personnel supposed to wear protective eyewear anytime they are handling chemicals?

Of course they are. However, remember that "incidents" do occur. The point is that when these incidents do occur needed emergency equipment must be available and in good working order. Ask yourself this question: If you contaminated yours eyes with chemical during an incident, how would you feel if the closest eyewash station did not function?

Talking about the closest eyewash fountain, it may surprise you to know that there are no fixed standards on the maximum acceptable distance of travel to reach an eyewash station. However, the National Safety Council has recommended that the travel distance should be no more than 25 feet or that the station take no longer than 15 seconds to reach. It is interesting to note that, in contrast to the National Safety Council's recommendation, the American National Standard Institute (ANSI) in its Z358.1-1990 standard "recommends" accessibility of emergency showers and/or eyewash stations within 10 seconds and placement at a distance no greater than 100 feet.

Again, it should be pointed out that these specifications are recommendations only. When it comes right down to it, common sense should dictate where the lab's eyewash fountains are to be placed. Remember, the effectiveness of any eyewash station is dependent on its accessibility. Placing eyewash stations on the other side of doorways is not making them accessible. The point is that no one should have to open a door or doors to reach an eyewash fountain. No one who is blinded and in pain should have to "feel" his/her way to a remote eyewash station. Depending on the size of the laboratory, eyewash stations should be located along a normal path of egress; for example, in a main laboratory corridor.

When in doubt as to where an eyewash fountain should be located the following standard rule of thumb on the topic may be of assistance.

Rule of Thumb: *The maximum time to reach an eyewash fountain should be determined by the potential effect of the chemical.* This thumb rule makes sense when you consider that if lab workers routinely work with very strong acids, then an emergency eyewash should be located within arm's reach, at their elbow, or no more than 10 ft in all directions (see Figure 13.1).

13.2.1 PERSONAL EYEWASH STATIONS

Simply stated, Personal eyewash stations are not suitable for use in the environmental laboratory. These personal eyewash stations usually are bottled and provided in buffered solutions and enclosed in emergency stations. Basically, there are two problems with personal eyewash stations: (1) they lack the volume of flow needed to properly flush the eyes of contaminant(s) and (2) the solution in the bottles may become contaminated.

In regards to volume of flow, eyewash fountains/stations should provide at least 0.4 gallons per minute at 25 psi or less. This eyewash pressurized-flow should be presented to the eyes in such a manner as to flood both eyes with aerated potable water. If the installed unit is intended to cover the entire face, then 3 gallons per minute is required according to ANSI 358.1.

Whatever style, type, or model of eyewash station chosen for use in the environmental laboratory, the main requirement is that the unit operate in the easiest manner possible. When a victim's eyes are contaminated with chemical, the last thing this victim should have to worry about is how to activate a complicated eyewash station.

13.2.2 WEATHER AND THE OUTDOOR EYEWASH/SHOWER STATION

Another point to keep in mind if you are planning on installing an eyewash station/fountain outdoors is the weather. Experience has shown that it is not unusual for a safety auditor to audit an industrial facility and find an outdoor eyewash/shower station that is inoperative because the pipes are frozen. It is important to ensure that outdoor stations are piped in such a manner that will ensure they are protected from the elements. Water at temperatures 60° to 70°F temperature range (normal tap water temperatures) should be used for emergency eyewash stations. *Note*: Lukewarm water at temperatures between 85° and 95° would be ideal.

13.3 EMERGENCY SAFETY (DELUGE) SHOWERS

In many installations emergency eyewash stations are combined with emergency showers to form one compact

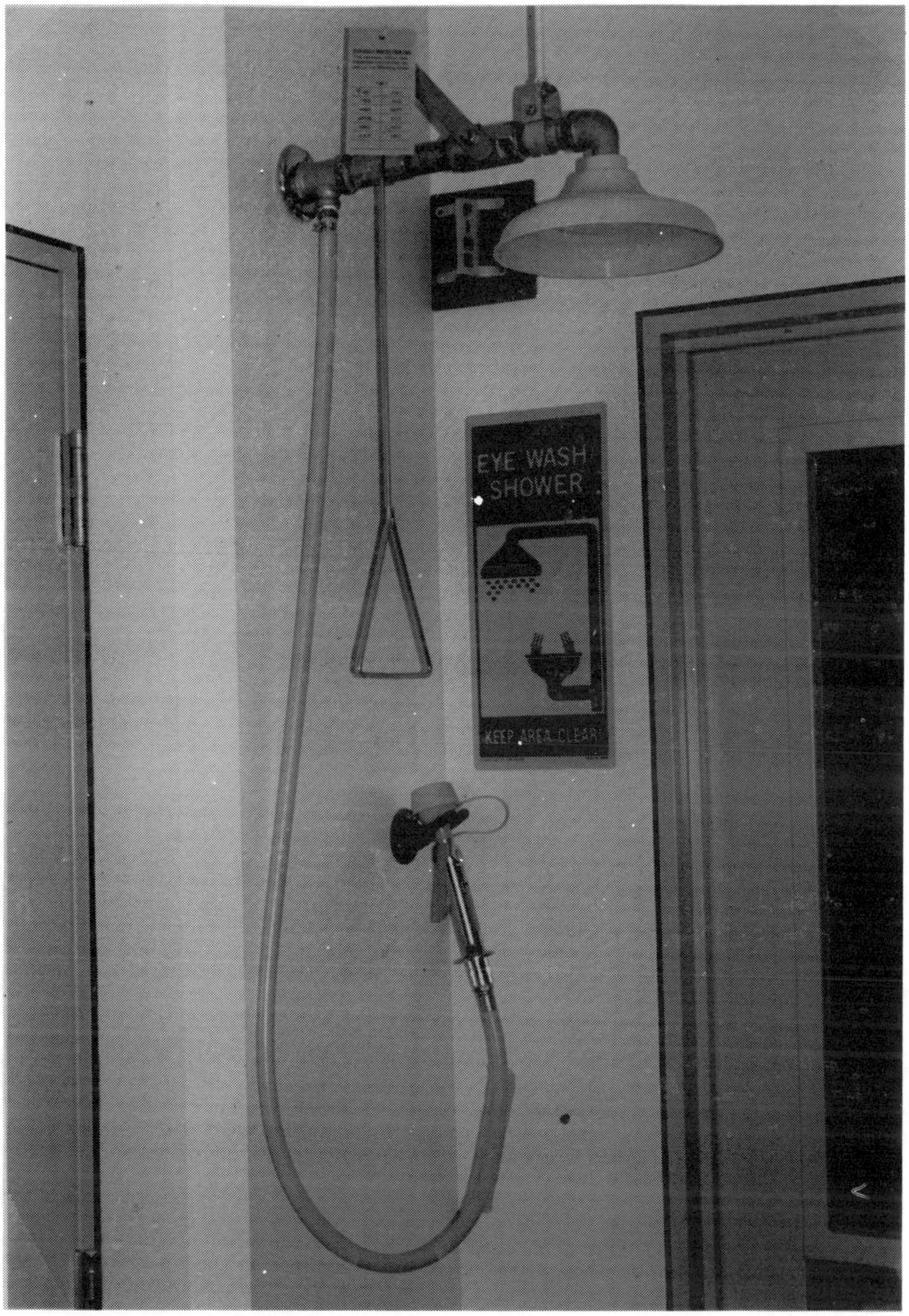

Figure 13.1. Emergency eyewash/shower station.

unit. Whether this is the case or not, the emergency shower is, like the eyewash station, standard safety equipment that must be installed in laboratory areas where chemicals are handled for immediate first aid treatment of chemical splashes and for extinguishing clothing fires. The lab's safety showers must have a quick-opening valve that will allow for drenching the victim immediately and must be large enough to accommodate more that one person if necessary; the quick-opening valve must be closed manually.

ANSI 358.1 recommends that safety showers be placed no more than 100 feet from any chemical handling area within the laboratory. Moreover, these safety showers are to be placed in such a way that there are on obstructions placed in front or near them.

Notwithstanding the importance of ANSI's standard 100 foot rule on location of emergency eyewash/showers, it should be obvious that when one's clothing is splashed by corrosive chemicals, having to travel 100 feet to get to the nearest safety shower would not be a pleasant undertaking. The point is that safety devices like safety eyewash/shower stations should be located in such a way so as to provide easy access to from all chemical handling areas within the laboratory. Remember, the first 15–20 seconds after an injury are critical, so emergency showers/eyewash stations should be placed as close to the hazard area as possible.

Along with properly locating the safety shower, it is important to ensure that the water supply feeding the shower is adequate. The water supply should be able to provide up to 30 gallons per minute. The temperature of the water supply to the safety shower is also important, but even more so than the temperature of eyewash station water. This is the case because the water from the deluge shower covers more of the body's surface area.

The safety shower should be easy to operate and should remain on until it is deliberately turned off. Some caution must be used when locating these shower stations. Obviously, it would not be wise to install a deluge shower next to an electrical power panel. Another point to keep in mind is that portable electrical equipment should not be moved near safety showers.

In summary, a few important points should be addressed here: (1) An important point concerning emergency eyewash stations and safety is that they should be ready for use at all times. These critical safety systems must be tested on a routine basis. Moreover, when found inoperable, they should be repaired or replaced immediately. Figure 13.2 shows an inspection tag used by HRSD to ensure that emergency eyewash/shower are inspected on a monthly basis. (2) According to ANSI, all emergency showers and eyewashes should be identified with highly visible signs and be located in well-lighted and highly visible areas. (3) As with any other safety device, work practice, or work procedure, worker training is vital.

The bottom-line is that the environmental laboratory can have all the safety equipment that it could possibly contain but if the workers are not properly trained on the use of the equipment, what good is it? Remember, any training conducted must be documented.

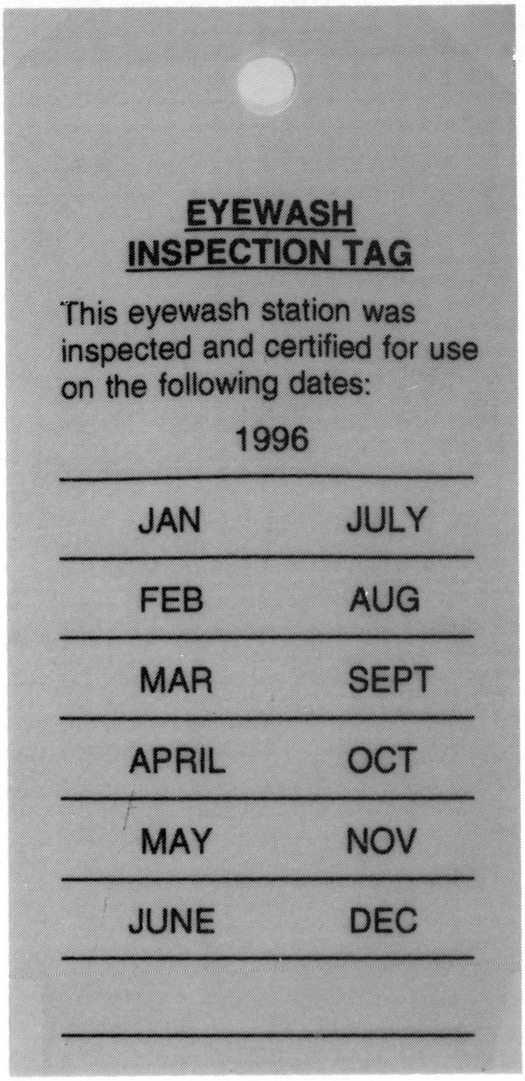

Figure 13.2. An example of an emergency eyewash station inspection tag.

CHAPTER 14

Respiratory Protection

The basic purpose of any respirator is, simply, to protect the respiratory system from inhalation of hazardous atmospheres. Respirators provide protection either by removing contaminants from the air before it is inhaled or by supplying an independent source of respirable air. The principal classifications of respirator types are based on these categories (NIOSH, 1987).

14.1 INTRODUCTION

According to the *Concise Oxford Dictionary*, a respirator is defined as an "apparatus worn over mouth and nose to warm or filter inhaled air or to prevent inhalation of poison gas." The use of respiratory protection by lab technicians in the environmental laboratory to filter out dusts, gases, vapors, mists and fumes is a common, everyday occurrence.

OSHA, under its 29 CFR 1910.134—Respiratory Protection Standard, requires that every employer protect workers from unacceptable levels of airborne hazards. While most air is safe to breathe, certain work operations and locations have characteristic problems of air contamination. Control measures are required to reduce airborne hazard concentrations to safe levels. When controls are not feasible, or while they are being implemented, workers must wear approved respiratory protection.

In order to comply with OSHA's Respiratory Protection Standard and to provide the necessary protection to workers, a respiratory protection program with the following minimum requirements is needed:

(1) Written operating procedures outlining the entire program
(2) Procedures for proper selection of respirators
(3) Procedures for training users in the proper use and limitations or respirators
(4) Cleaning, maintenance, and storage procedures
(5) Inspection
(6) Medical surveillance for respirator users

The environmental lab's written Respiratory Protection Program should be an organized approach that will assure employees a safe workplace by providing specific requirements in such areas as listed in the following:

(1) Designation of individual responsibilities
(2) Definition of various terms used in the Respiratory Protection Program
(3) Designations of types of respirators and their applications and limitations
(4) Designation of procedures for respirator selection and distribution
(5) Designation of procedures to be used for inspection and maintenance of respirators
(6) Designation of procedures for employee respirator fit-testing
(7) Designation of a procedure for medical surveillance
(8) Designation of a training program for lab personnel participating in the Respiratory Protection Program
(9) Documentation procedures for lab personnel participating in the Respiratory Protection Program

The Chemical Hygiene Officer (CHO) is responsible for the laboratory's Respiratory Protection Program. Specifically, the CHO should perform the tasks listed in the following:

(1) Develop and modify, as necessary, the lab's Respiratory Protection Program.
(2) Check and review, quarterly, all respirator inspection records.
(3) Compile and maintain a master respirator inventory list for the lab.
(4) Implement an ongoing respirator training program.

(5) Conduct initial and annual employee respirator fit-testing.
(6) Provide initial and annual spirometric evaluation to ensure that employees are physically capable of wearing a respirator under their given work conditions.
(7) Provide technical assistance in determining the need for respirators and in the selection of appropriate types of respirators.
(8) Ensure copies of respirator training records, fit test results, spirometric results, and medical examination results are placed in each worker's personnel/medical record.

14.2 TYPES OF RESPIRATORS

Respirators are generally classified into three types: chemical cartridge, air line respirators and self-contained breathing apparatus (SCBA).

14.2.1 CHEMICAL CARTRIDGE RESPIRATORS

Chemical Cartridge Respirators are usually considered to be low-capacity gas masks. They consist of a facepiece, which fits over the nose and mouth of the wearer. Attached directly to the facepiece is a small replaceable filter-chemical cartridge. This type of respiratory protection is usually used where there is exposure to solvent vapors or dust and particulate matter as with sandblasting, spray coating, and/or degreasing. This type of respirator is not to be worn in an Immediately Dangerous to Life and Health (IDLH) atmosphere.

14.2.2 AIR-LINE (SUPPLIED AIR) RESPIRATORS

Air-line respirators may be of the helmet, hood, or mask type. Moreover, they may be fed class D breathing air (normal air with contaminants removed) via a Cascade-Bottle system or by mechanical compressor. This device provides air to the wearer through a small-diameter, high-pressure, hose line from a source of uncontaminated breathing air. The source is usually derived from a compressed air line with a reducing valve in the hose to reduce the pressure. When used with a reciprocating engine-driven compressor system, a filter must be included in the system (between the compressed air line and the respirator) to remove oil and water mists, oil vapors, and any particulate matter that may be present in the compressed air. In addition, compressors that are internally lubricated require precautionary detection and warning equipment installed to detect and warn against the presence of carbon monoxide. Where the air supply for air-line respirators is taken from the compressed air line, a carbon monoxide alarm must be installed in the air supply system. It is important to point out that an operational check should be made on this carbon monoxide alarm system prior to its use. Air-line or supplied air respirators have limited use in the environmental laboratory; however, they are used extensively in taking environmental samples, e.g., asbestos, in the field.

14.2.3 SELF-CONTAINED BREATING APPARATUS (SCBA)

This type of respirator provides Grade D breathing air either from compressed air or breathing air cylinders or by chemical action in the chemical canister attached to the apparatus and enables the wearer to be independent of any outside source of air. This equipment may be operable for periods between one-half to two hour (depending on cylinder size and usage). The operation of the self-contained breathing apparatus is fairly complex, and it is therefore necessary that the wearer have special training before being permitted to use it in an emergency situation.

Because the oxygen-producing mechanism is self-contained in the apparatus, it is the only type of equipment that provides complete protection and at the same time permits the wearer to travel for considerable distances from a source of respirable air. SCBA's can be used in most industrial applications.

14.3 RESPIRATOR SELECTION AND DISTRIBUTION PROCEDURES

Respirators used in the environmental laboratory should be selected by the Chemical Hygiene Officer with input from the chemist, biologist, or technician who will be wearing the respirator. To assist in the selection of the correct cartridge to be used for various atmospheric contaminants a reference such as Table 14.1 should be used. From Table 14.1 it can be seen that cartridge selection is based on matching the proper color-coded cartridge with the type of protection desired. Selection is also dependent upon, of course, the quality of fit and the nature of the work being done.

Cartridge type respirators are usually issued to the workers who are required to use them. Each of these individually assigned respirators is identified in a way that does not interfere with its performance. In the laboratory, questions related to the respirator and cartridge selection process should be referred to the CHO.

14.4 RESPIRATOR INSPECTION, MAINTENANCE, CLEANING AND STORAGE

To retain their original effectiveness, respirators should be periodically inspected, maintained, cleaned and properly stored. Moreover, when respirators are assigned to

TABLE 14.1. Respirator Cartridge Color-Coding System (partial listing).

Atmospheric Contaminant to Be Protected Against	Colors Assigned
Acid gases	White
Chlorine	White/yellow
Organic vapors	Black
Ammonia gas	Green
Acid gas and organic vapors	Green/white Stripe
Carbon monoxide	Blue
Acid, gases, organic vapors and ammonia gases	Brown
Particulates (dusts, fumes, mists, for, or smoke)	1/2-inch gray stripe around normal canister color
All the above contaminants	Red with 1/2-inch gray stripe
Asbestos	Pink (HEPA)

laboratory workers for their exclusive use, the worker is responsible for routine maintenance and care of the respirator. In addition, when the laboratory institutes its written Respiratory Protection Program, it must stipulate a respirator inspection protocol. The point is that in addition to a detailed respirator inspection protocol with accompanying recordkeeping, this respirator inspection format should include self-inspections to be performed by each worker who is issued a respirator.

14.4.1 RESPIRATOR INSPECTION

Respirators should be inspected before and after each use, after cleaning, and whenever cartridges or cylinders are changed. Appropriate entries should be made in workcenter respirator "Inspection After Each Use" records or similar documentation. If, for example, a respirator is taken out of service for any reason, this should be noted in the workcenter respirator records. When this same respirator is put back into service it must be thoroughly inspected to ensure it is safe for use. Respirators that have been designated for general or emergency use should be inspected at least monthly and cleaned and disinfected after each use. The CHO is responsible for ensuring that the lab's respirators are inspected on a routine basis.

14.4.2 RESPIRATOR MAINTENANCE

Respirators that are found to be deficient for any reason during routine inspections must be replaced or repaired prior to use. Keep in mind that in-house respirator repairs should be limited to changing of cartridges, canisters, cylinders, filters, head straps and those items as recommended by the manufacturer. *No attempt* should be made to replace components, or make adjustments, modifications or repairs beyond the manufacturer's recommendations.

14.4.3 RESPIRATOR CLEANING

Laboratory workers assigned respirators should be instructed on the proper procedure for cleaning respirators. Respirator manufacturers specify these correct cleaning procedures. Usually, they recommend that the user break down the respirator into its component parts and wash each part thoroughly with a mild detergent. After rinsing and drying the respirator, it should be reassembled and then closely inspected to ensure it is fit for service. During the inspection of a cartridge type respirator it is important to pay attention to the inhalation and exhalation valves, the condition of the facepiece and the elasticity of the straps. If the purifying elements need replacement, this is a good time to do so. The final step should include sanitizing the respirator with an approved disinfectant pad. After inspection, cleaning and necessary repairs, the respirator should be sealed in a clear plastic bag and stored in a cool dry area to protect against dust, sunlight, extreme heat, extreme cold, excessive moisture or damaging chemicals.

14.5 RESPIRATOR FIT-TESTING PROCEDURES

In its Respiratory Protection Program, the environmental laboratory should include a fit-test protocol for all workers required to wear negative-pressure respirators. The goal of this fit testing procedure is (1) to provide the worker with a face seal on a respirator that exhibits the most protective and comfortable fit and (2) to instruct the worker on the proper uses of respirators and their limitations.

Generally, there are three levels of respirator fit-testing: initial, annual, and pre-use self-testing.

The initial and annual fit test are rigorous procedures used to determine whether the worker can safely wear a respirator. These tests are usually conducted by the organization's safety professional, industrial hygienist, or person specially trained in this area (see Figure 14.1). The fit-testing procedure requires special equipment and test chemicals such as banana oil, irritant smoke or saccharin. In general, any change to the face or mouth may alter respirator fit; if this is the case, the safety professional should be requested to perform a new fit test. Upon completion of initial and annual fit-testing, a record of the test should be documented and included in the worker's training or medical file.

(a)

(b)

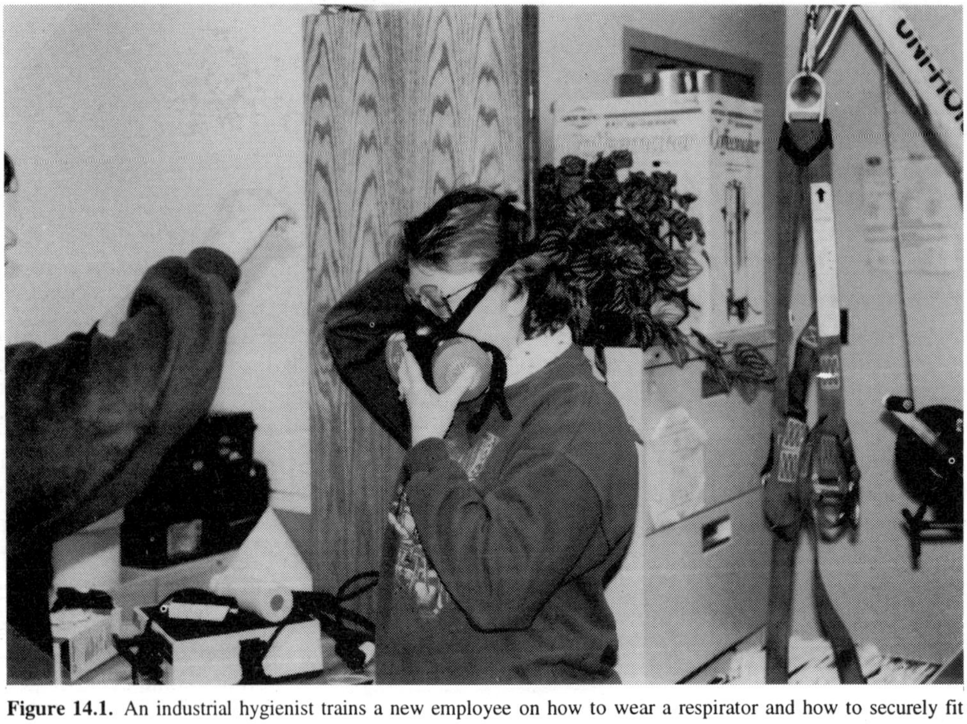

Figure 14.1. An industrial hygienist trains a new employee on how to wear a respirator and how to securely fit the respirator. Then the employee's respirator is tested with irritant smoke to ensure proper fit.

(c)

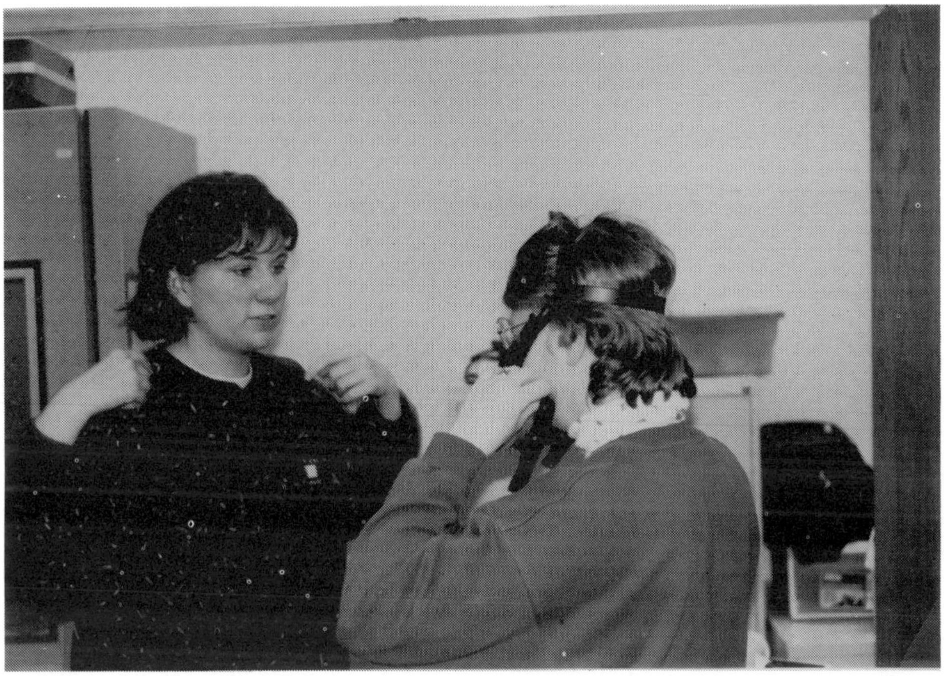

(d)

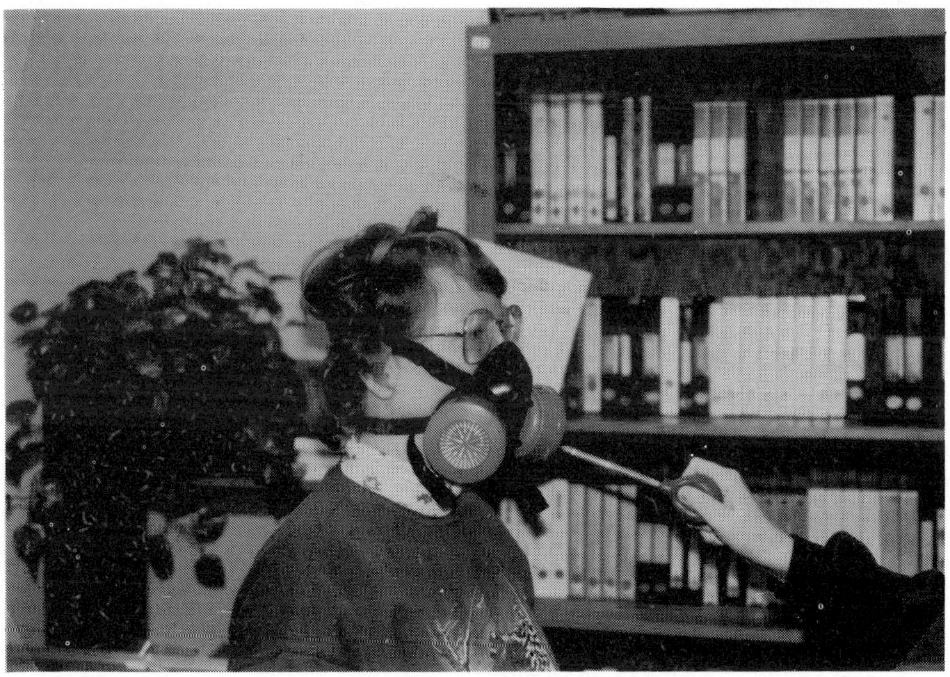

Figure 14.1. *Continued*

Pre-use self testing is a routine requirement for all workers who wear respirators. Each time the respirator is used, it must be checked for positive and negative seal. To conduct a positive pressure check the respirator is placed on the face and straps are adjusted for firm but comfortable tension. The user then places palm of hand over the respirator's exhalation valve. He or she then takes a deep breath and gently exhales a small amount of air. Holding his/her breathe for 10 seconds, the mask should feel as if it wants to pop away from the face, but no leakage will occur.

In conducting the negative pressure check the user, still wearing the respirator, covers both filter cartridges with the palms, and inhales slightly to partially collapse the mask. This negative pressure should be maintained for 10 seconds. If no air leaks into the mask, it can be assumed the mask is fitting properly.

There are several factors that may affect the fit and protection afforded by a respirator. One such factor is the presence of facial hair between the mask and the surface of the face. Facial hair will permit the passage of unpurified air into the interior of the facepiece. While OSHA does not specifically state that facial hair is not allowed for respirator users, it does state that the user must be able to seal the respirator to his/her face. Thus, if one has facial hair and a good seal is not possible between the face and respirator because of the facial hair, then the facial hair must be removed or trimmed to facilitate a tight fit.

Another factor that can have a bearing on how well a respirator will fit the face of the user is the face itself. Any structural damage to the face (e.g., from broken facial bones received in an accident) could make the fit difficult to achieve.

Workers who wear corrective eyewear pose another respirator fit problem. While it is not the responsibility for the employer to purchase and provide special prescriptive lenses for the user, it is the responsibility of the employer to ensure that the worker is able to see while wearing a respirator. To solve this problem the user is usually required to special order and pay for correction lens that can be mounted inside full facepieces. The facepieces and lens must be fitted by qualified personnel to provide good vision, comfort and an air-tight seal.

14.6 MEDICAL SURVEILLANCE

OSHA is clear in its statement stipulating that no one should be assigned a task requiring the use of respirators who is not certified physically able to do the work while wearing the respirator. The point is that the use of a respirator can impose significant physiological and psychological stress on the wearer. Since such stress may place the user at an increased risk of illness or injury, a licensed physician should evaluate the fitness of each user assigned to wear a respirator.

At the very least this medical evaluation should include:

- medical history
- visual acuity
- hearing acuity
- facial abnormalities
- pulmonary function test (see Figure 14.2)
- cardiovascular fitness
- psychological tests
- exercise stress test

After the evaluation, it is important to ensure that the physician provide a written statement of the results of the examination, including whether or not the individual is medically qualified to use a respirator. In some cases, the physician may qualify an individual to use a respirator on a limited basis. These limitations will vary from no use other than for emergency escape purposes to restrictions on length of time the respirator is to be worn.

14.7 TRAINING

No worker should wear a respirator before medical evaluation, fit-testing, and training have been completed. Generally, organizational safety professionals perform respirator training (see Figure 14.1). This training, however, can be conducted by any qualified person. At the minimum, the environmental laboratory should provide training on initial assignment, whenever the potential for exposure changes, and at least annually. This training should provide information with regard to:

- preuse inspection
- functional components of respirator
- donning instructions
- air-purifying cartridge/canister selection
- limitations
- positive/negative-pressure checks
- emergency instructions
- care and maintenance
- cleaning instructions
- storage requirements

As with any type of organizational safety training, the training must be documented and maintained on file in the worker's training or medical record.

14.8 RECORDKEEPING REQUIREMENTS

As stated above, recordkeeping and documentation are an important part of the lab's respiratory protection

Figure 14.2. A new employer is given a pulmonary function test with a spirometer. This test screens the employee for suitability to wear a respirator on the job.

program. Moreover, OSHA, during an audit, will want to review all documentation related to respiratory protection including documents that will verify (1) training; (2) medical evaluations with physician's written statement of approval or non-approval to wear respirators (if limitations apply, they must be stated); (3) fit-testing; (4) care and maintenance procedures; (5) a written respiratory protection program; and (6) standard operating instructions or safe work practices for specific routine respirator use.

PART VI

EMERGENCY RESPONSE

CHAPTER 15

Emergency Response

> Having an emergency plan will not mean that a disaster will not happen, but with a plan in place, the chances of an accident becoming a disaster are greatly diminished. An emergency plan should be constantly reevaluated and updated to ensure that changing conditions and personnel are taken into account. An obsolete plan is little better than no plan at all. (Newton, 1987, p. 3)

15.1 INTRODUCTION

THE very nature of working in an environmental laboratory with hazardous chemicals makes emergencies very real possibilities. These emergencies strike immediately and unexpectedly and require immediate response. Although it is the accidents with hazardous materials that are most likely to lead to emergency response situations in the environmental laboratory, other possible emergency situations must also be taken into account. For example, fires, storms (hurricanes and tornadoes), earthquakes, electrical shock, power failure, and medical emergencies are real possibilities [these events may be related to (in the case of fire and medical problems) or compounded by hazardous materials but not in every situation].

This chapter addresses emergency response procedures used to control and contain losses of property and prevent injury to workers in the environmental laboratory due to fire, explosion, toxic chemical spill, and/or accidents. Moreover, an emergency response procedure, currently used in an environmental laboratory, is presented as an example of the type of procedure that is required. Along with standard emergency response procedures for use in the lab, several work practices related to specific laboratory equipment is included. Finally, emergency medical response procedures for the environmental laboratory are discussed.

15.2 EMERGENCY RESPONSE: REQUIREMENTS

When a hazardous materials (hazmat) incident occurs, decisive action is immediately required. Decisive action, however, cannot be taken in a safe and sound manner unless those reacting follow a plan, a routine, a protocol that will allow them to mitigate the situation with the least degree of danger to themselves, others, and the environment. Because of the need for decisive action and a written plan of action for hazardous materials incidents that may occur in the laboratory, OSHA has mandated two basic but far-reaching requirements: (1) Every laboratory is responsible for having a comprehensive emergency action plan that complies with OSHA's Standard on Emergency Plans and Fire Prevention Plans (29 CFR 1910.38). This emergency action plan must be ready and available for implementation during emergency events. (2) OSHA also has a special regulation dealing specifically with spills of hazardous chemicals. This regulation, called the Hazardous Waste Operations and Emergency Response (HAZWOPER) standard, is found at 29 CFR 1910.120 and has direct applicability to the environmental laboratory where hazardous chemicals are used.

At the heart of safe work practices with hazardous materials handling is good emergency planning. Hazardous materials mishaps are always characterized by their potential for complexity (Hall, 1994). This is why OSHA specifically mandates emergency planning. The environmental laboratory's emergency plan should address actions to be taken in case of fire and/or explosion and should include, at a minimum, the following:

- a list of emergency numbers to be used to alert lab personnel, local police, fire, emergency rescue, and medical personnel
- the preferred means of reporting fire and emergencies

Figure 15.1. Material Safety Data Sheet (MSDS) container. This container is located in the entranceway to an environmental laboratory; it is specifically located here to aid emergency response personnel in determining the hazardous materials contained within the lab. Obviously, this is a good idea.

- names and phone numbers of personnel to be contact in an emergency
- rescue and medical duties for those who are designated to perform them
- emergency escape procedures and emergency escape route assignments
- procedures to account for all employees after an emergency evacuation
- procedures to be followed by employees who remain to operate or shut down operations before they evacuate
- Material Safety Data Sheets (MSDSs) detailing the chemical and physical properties of hazardous materials stored or handled in the laboratory, and specific emergency response measures to be implemented to contain these materials in the event of an emergency (see Figure 15.1)

Even with a written plan it is important to keep in mind that under no circumstances should untrained personnel respond to a hazardous materials incident.

15.3 EMERGENCY RESPONSE PROCEDURES FOR THE ENVIRONMENTAL LABORATORY (AN EXAMPLE)

The following protocol for Emergency Response Procedures in the Environmental Laboratory is an example that has been used for years. This procedure is beneficial because it has proven its worth; it has been tested.

15.3.1 CLEANING UP HAZARDOUS MATERIALS SPILLS

1. ***Slightly Hazardous Materials (Weak Acids, Bases, Oils, etc.)***

Procedure

- Absorb using paper towels, shaker dispensed Spill-X-A type for acids and Spill-X-C type for caustic or basic substances, spill-control pillows.
- Transfer to a plastic bag, place in trash container for solid wastes. Clean floor (or other spill

surface) thoroughly with with soap and water; dry surface completely to prevent slips or falls.

2. ***Mercury Spill***

Procedure

- Contain the spill in a small area and warn co-workers not to walk on spill area.
- Use gloves and remove any sources of heat that might volatilize the mercury. Do not use an ordinary vacuum cleaner; it will aerosolize and spread the mercury all over the laboratory.
- If the mercury has reached a porous surface (floor, lab bench, etc), then special chemical inactivators must be used. Hg-X is mixed with water and the contaminated area wiped with the mixture. Zinc powder mixed with 5% to 10% sulfuric acid to form a paste can be spread over the contaminated area and allowed to dry. Both chemicals concentrate the mercury to facilitate its removal.
- After the materials have been removed, check the area for the presence of mercury using a "mercury sniffer."
- Several cleanings may be necessary to remove all of the mercury.

3. ***Concentrated Acids/Bases***

Procedure

- Wear heavy gloves that are acid resistant, goggles and body protection (lab coat or rubber apron).
- Take care not to step on the spilled materials.
- Absorb the spill using acid or base Spill-X or absorbent pillows.
- Products designed to neutralize acids or bases can produce large amounts of heat and may splatter. Treat the spill in small sections.
- Sweep or scoop the neutralized materials into bags and dispose of in the trash (if neutralized).
- Clean area thoroughly with soap and water; dry thoroughly.

4. ***Flammable/Volatile Chemicals***

Procedure

- Immediately extinguish all open flames.
- Shut down all sources of ignition (sparking equipment, incubators, flame AA, etc.).
- Remember, for extremely explosive gases, the spark from turning on a light switch can be sufficient to ignite the gas.
- Vapors can travel from open containers or other sources of volatile chemicals. Ethanol, methanol, propanol, and acetone, for example, should be isolated within a fume hood when in an open container. Special flammable storage cabinets should be used to store these materials.
- Alert other workers present in the lab of the situation and seek assistance if necessary.
- Open any windows/doors and the fume hood sash to help dissipate vapor.
- Absorb the spill using spill-control pillows or other absorbent materials designed for volatile agents.
- Place the saturated pillow or absorbent materials into a plastic bag (hazwaste bag) and place the bag into a fume hood or outside the main laboratory, if possible.
- Use gloves and shovels, scoops or brooms to move the material. (All materials used to transport the spilled substance must be decontaminated or, if not possible to decontaminate, disposed of together with the hazardous substance, as with mercury).
- Return to the site and, after the remaining material has evaporated, clean up the contaminated area using soap and water if possible.
- If cleanup procedures are in doubt, mark the contaminated area with a warning sign and isolate with hazard tape, cones, or other type barricade material. Immediately report the spill to the lab supervisor.

15.3.2 EXPOSURE TO CHEMICALS

(1) If a chemical comes in contact with unprotected hand(s), wash hands immediately using warm water and a strong soap such as dishwashing detergent. *Do not use* alcohol, acetone, or other solvents (these will facilitate absorption of the chemical through the skin).

(2) Splashes into eyes, nose, or lip area must be removed immediately using an eyewash station/hose or by placing face under a steady stream of lukewarm running water. A mild soap may be used for the lips and nose, but only running water is to be used for the eyes. It is important that the eyes be held open in the stream of water and that contact lenses be removed to facilitate cleansing of the eyes. Roll the affected eye during the rinsing, which should continue for at least 15 minutes. Seek medical attention immediately.

(3) If a chemical has been spilled onto clothing or other parts of the body, remove any contaminated clothing or jewelry immediately and wash the body part with warm water and soap for at least 10 minutes. Contact with phenol, DMSO-containing solutions, cyanate or pyridine is especially serious, since these compounds pass through the skin and into the bloodstream rapidly. Dispose of the contaminated clothing in the

same manner that the chemical wastes are treated. Seek medical attention.

(4) If a toxic substance is spilled and inhaled, the substance is absorbed quite rapidly into the bloodstream. Those personnel exposed in such a manner should leave the contaminated area and seek immediate medical attention. Attempt to cough up as much bodily secretions as possible and rinse mouth with water. Remove stunned or unconscious victims from the contaminated area only with the use of SCBA (self-contained breathing apparatus). Never attempt to remove a victim from a room containing toxic vapors without using protective breathing apparatus. Be cautious about using CPR on a person who has inhaled toxic vapors—in order not to inhale any of the poison from the victim. Administer emergency oxygen, if trained to do so. Place warning signs at the room entrance where the accident occurred to indicate contaminated area.

(5) If materials have been ingested, rinse mouth with warm water repeatedly. If corrosives where ingested, do not attempt to neutralize the chemical, since this will only generate heat that may further damage sensitive tissues. Most physicians recommend diluting the corrosive materials using large quantities of water or milk. Do not attempt to induce vomiting if corrosives have been swallowed since this may rupture the esophagus. If other toxic substances have been swallowed, drink large amounts of water or milk to dilute the toxin. Vomiting may be induced only if corrosives or hydrocarbons (petroleum distillates) have not been swallowed. Seek medical attention and be prepared to inform the medical authorities about the nature and approximate amount of chemical that was ingested.

15.3.3 FIRE SAFETY

Preventing Fires

A fire may be prevented or stopped by removing any of the three necessary elements (fuel, oxygen, and an ignition source) and interfering with the chain reaction that occurs in ignition.

(1) ***Restrict the fuel*** by using only minimal amounts of flammable materials in the work area. Large containers of flammable solvents of other chemicals should be kept in an approved safety cabinet with doors closed.

(2) ***Limit access to oxygen*** by keeping all containers capped when not in use. Especially with volatile solvents it is important to ensure that the containers are closed immediately after dispensing the necessary amounts. Remove the working container from the work area after it is no longer needed.

(3) ***Keep ignition sources away from available fuels.*** Although it should be obvious that an ignition source never should be near combustible materials, it is often overlooked. It may not be uncommon to use solvents in close proximity to electrically powered equipment such as stirrers, ultrasonic baths, or even ovens. Since solvent vapors may travel some distance, a spark or heating coil may ignite the vapor phase, which may then serve as a conduit back to the fluid phase. One should never store solvents in a standard refrigerator since a spark from the thermostat, compressor, or other electrical relays may cause an explosion. Instead, a specially designed explosion-proof or flammables refrigerator may be used to store volatile chemicals. Self-defrosting refrigerators cannot be safely modified to store volatile materials.

Note: Care must be taken to ensure that no open solvents are left near the high-temperature ovens.

Stopping Fires

(1) Evacuate all persons to a safe location. Evacuation is *always* to an upwind direction of fire or chemical spill. Note the wind direction and assemble in the designated mustering location on the side of the environmental laboratory building that is upwind. If the upwind side is the street, gather directly across the street. Everyone must gather at the pre-arranged assembly point. It is important in order to determine if anyone is still in the building. The Chemist for each section will perform a head-count to assure that all technicians working that day are safe. If the Chemist is not present, the senior technician in the section will count those present. If anyone is injured, do not attempt to move them unless the fire appears to be out of control.

(2) If it is possible to extinguish the fire safely, do so immediately. After determining the type of fire, select the proper fire extinguisher and position yourself between the fire and the room exit. Do not permit the fire to block your exit. After the fire is safely out, notify the fire department so that it may evaluate the need for further measures.

(3) If it is not possible to quickly extinguish the fire, close off the room and activate the fire alarm to alert others in the building. Telephone the fire department, giving the exact location of the fire. Indicate the nature of the fire (electrical, chemical, flammable liquids, etc.) and if hazardous materials are involved. Proceed to the assembly point and be prepared to meet the fire department with further directions as needed.

Classes of Fires and Proper Extinguishers

(1) ***Class A fires:*** This is the most common type of fire; wood, cloth, paper, or other common combustibles are involved. This type of fire is best extinguished with a pressurized tank usually containing either plain water or water in combination with special wetting agents and aqueous film-forming foam (AFFF). This type of extinguisher is often found in corridors and in offices. A special fire blanket may also be used.

(2) ***Class B fires:*** Flammable liquids are involved. These fires are extinguished by using pressurized carbon dioxide, halogenated compounds such as Halon, dry chemicals (ammonium phosphate, sodium/potassium bicarbonates, potassium chloride), or possibly by means of AFFF.

(3) ***Class C fires:*** Live electrical circuits are involved either as the cause of the fire or as simply being near the fire. Use carbon dioxide, Halon, or dry chemical extinguishers. Aqueous-based or electrically conductive reagents must not be used to fight such fires.

(4) ***Class D fires:*** Combustible metals, such as magnesium, sodium or potassium are involved. Class D Fire Extinguishers contain dry chemicals such as sodium chloride containing a thermoplastic binder that forms a solid suffocating crust over the fire. None of the common extinguishers should be used since they can increase the intensity of the fire by adding an additional chemical reaction.

Using Fire Extinguishers

Although fire is generally a panic situation, try not to panic—when panic sets in, mistakes are made and a bad situation turns worse, usually in an exponential fashion. To fight the fire, direct the spray at the base of the fire, not at the flames. The object is to suffocate and cool the chain reaction that is occurring at the level of the fuel. Slowly sweep the extinguishing agent over the base of the fire until all flames are extinguished. Even after the flames have subsided, continue to apply the agent to prevent a flare-up or flashback of the fire.

Note: The following hazards/hazard categories/ procedures may or may not be related to emergency response in the laboratory; however, each of these hazards is important to consider in any laboratory safety program. Lab technicians and other lab personnel need to be thoroughly aware and familiar with the subject areas that follow.

15.3.4 ELECTRICAL SAFETY

(1) Keep electrical equipment away from lab sinks and running water. In addition, wet floors should not only be mopped to prevent a slip hazard but also to prevent electrical shock. A rubber mat provides some protection against electrical shock, but should not be depended on as a substitute for proper grounding and bonding of electrical equipment and maintaining floors in a dry condition.

(2) Check electrical cords for fraying; never allow them to dangle over sinks. Report damaged cords to the electricians immediately, and label the equipment (tag it out) so that it will not be used.

(3) Any electrical equipment that must be near a sink must be secured in place and isolated from splash or spray.

(4) Outlets on the backsplash of a sink must be equipped with a ground-fault circuit interrupter (GFCI). Rubber mats should be placed in front of sinks that have electrical outlets on the backsplash.

(5) Do not handle electrical equipment with wet hands. Keep areas surrounding sinks dry. Do not touch an electrical device and the sink, faucet, or water at the same time. Do not reach into a sink to retrieve an electrical item that has dropped in. Either unplug the item or cut off the circuit breaker (safest) and then unplug, before retrieving. Label such equipment for evaluation by the electrician before using it again.

(6) Equipment that generates or uses large amounts of current or magnetic fields should only be operated by properly trained individuals, following the manufacturer's recommendations. Always unplug equipment when cleaning or servicing.

(7) All electrical equipment should use a three-prong or other properly grounded connector. Do not use two-to-three prong adapters without ascertaining that a proper ground has been established (check with an electrician). Determine the location and how to use circuit breakers or fuse boards to cut off the power to equipment in an emergency.

(8) Servicing of electrical equipment should be left to qualified personnel only. Disconnect the equipment, tag it out, and notify the supervisor.

15.3.5 FIRST AID FOR SHOCK VICTIMS

If contact with a live electrical circuit has taken place, the first step should be to cut off the current at the circuit breaker box. If this is not possible, separate the victim from the electrical source using a non-conductive material such as a lab coat, wooden chair, or rubber hose. Make sure that you do not become involved in the circuit (avoid water spills or making contact with grounded objects such as water faucets or other electrical equipment). If victims are unconscious, attempt to arouse them by gentle shaking. If breathing has stopped or a pulse cannot be felt, apply cardiopulmonary resuscitation (CPR) following approved methods. Send someone to call for medical help.

15.3.6 PHYSICAL AND MECHANICAL HAZARDS

Sharp Objects

- Keep glass shards or cracked glass vessels in a thick-walled box lined with a plastic bag. When filled, tape the box shut and dispose of in the appropriate manner. (If you are wondering why this text puts so much emphasis safety in the use of glassware, maybe Figure 15.2 will help to explain this emphasis).

Vacuum Pumps

- Pumps become very hot during operation. Be cautious of burn hazards.
- Used pump oil should be disposed of in the same fashion as other used oil products, except when the pump has been used to vacuum aspirate toxic or carcinogenic substances. Used oil from these pumps is hazardous waste, and must be disposed of accordingly. This kind of pump should not be vented into the work area. Exhaust the pump into a fume hood or outside the building.

Ovens

- Never place volatile organic solvents such as acetone, alcohol, toluene, etc., into an oven. Vent ovens to fume hoods.

Cryogenic Gases

- Frostbite hazard exists when working with gases such as liquid helium, nitrogen, and carbon dioxide, resulting in serious burns. Use insulated gloves and a face shield when dispensing cryogens. Sandals or open-toed footwear must not be worn. Non-absorbent laboratory clothing is recommended. If liquid cryogens are spilled onto ordinary laboratory clothing, it must be removed as rapidly as possible, since it may freeze onto the skin.
- If liquid carbon dioxide is to be used, take care that the dry ice does not block the outlets and gas lines.
- Cryogens should not be used in confined spaces, where accidental release will displace oxygen.

15.3.7 COMPRESSED GAS SAFETY

"The very properties that make compressed gases, liquefied compressed gases, and cryogenic liquids useful in almost every area of modern life can also make them dangerous when mishandled" (Compressed Gas Association, 1990).

"Dangerous when mishandled" is the key phrase. Fortunately, years of experience with compressed gas has resulted in safe work practices and equipment which, if properly used, results in complete safety.

Compressed gases are usually stored in cylinders or other containers at high pressures and must always be handled with care. While compressed gas containers are designed to meet rigid regulatory specifications and are safe for their intended use, they can be extremely hazardous if abused or misused. Thus, care and caution is always advised when working with or around compressed gases.

Compressed Gas Cylinder—A Description

A compressed gas cylinder may have a series of numbers stamped into it. For instance, the designation "DOT 3AA 2300" refers to Department of Transportation (DOT) specification of tank for 2,300 psi at 70°F. Directly under this designation may be a serial number such as "Z48976" to identify the specific tank. Other identifiers, such as "10 PB 89+", indicate the most recent inspection date or date of manufacture. In this case, an inspector with initials PB inspected the cylinder in October, 1989. The plus sign reveals that this tank may be filled to 10% over the 2,300 psi service pressure. In addition, the tank must be clearly marked as to the contents and whether or not the gas is hazardous.

All compressed gas cylinders are equipped with a safety device that permit gas to escape from the cylinder in the event that excessive pressures develop inside the tank. A hand wheel is used to open or close the valve, but not to adjust the pressure of the escaping gas. An outlet connection is provided for attaching a pressure regulator valve to the main cylinder valve. The threads of these connections are designed to accommodate special regulators or attachments and to prevent the connection of improper accessories to the tank (e.g., the wrong type of regulator). A two-stage pressure regulator valve is normally used with the standard compressed gas cylinder. The gauge nearest the cylinder indicates the pressure inside of the main tank, while the second gauge indicates the discharge pressure of the gas. The discharge pressure is regulated by the handle on the large portion of the regulator. In most of these valves, the discharge pressure may be increased by turning the valve clockwise, rather than counterclockwise as one might expect. A final delivery valve may be present to turn the flow of gas from the regulator on or off. It has no effect on the dispensing pressure, but only on the rate of discharge. All tanks have a cylinder cap that is screwed over the cylinder valve to protect it when the tank is not in use or during shipping. *This cap must be securely screwed in place on all tanks that do not have a regulator or on tanks that are not in use.*

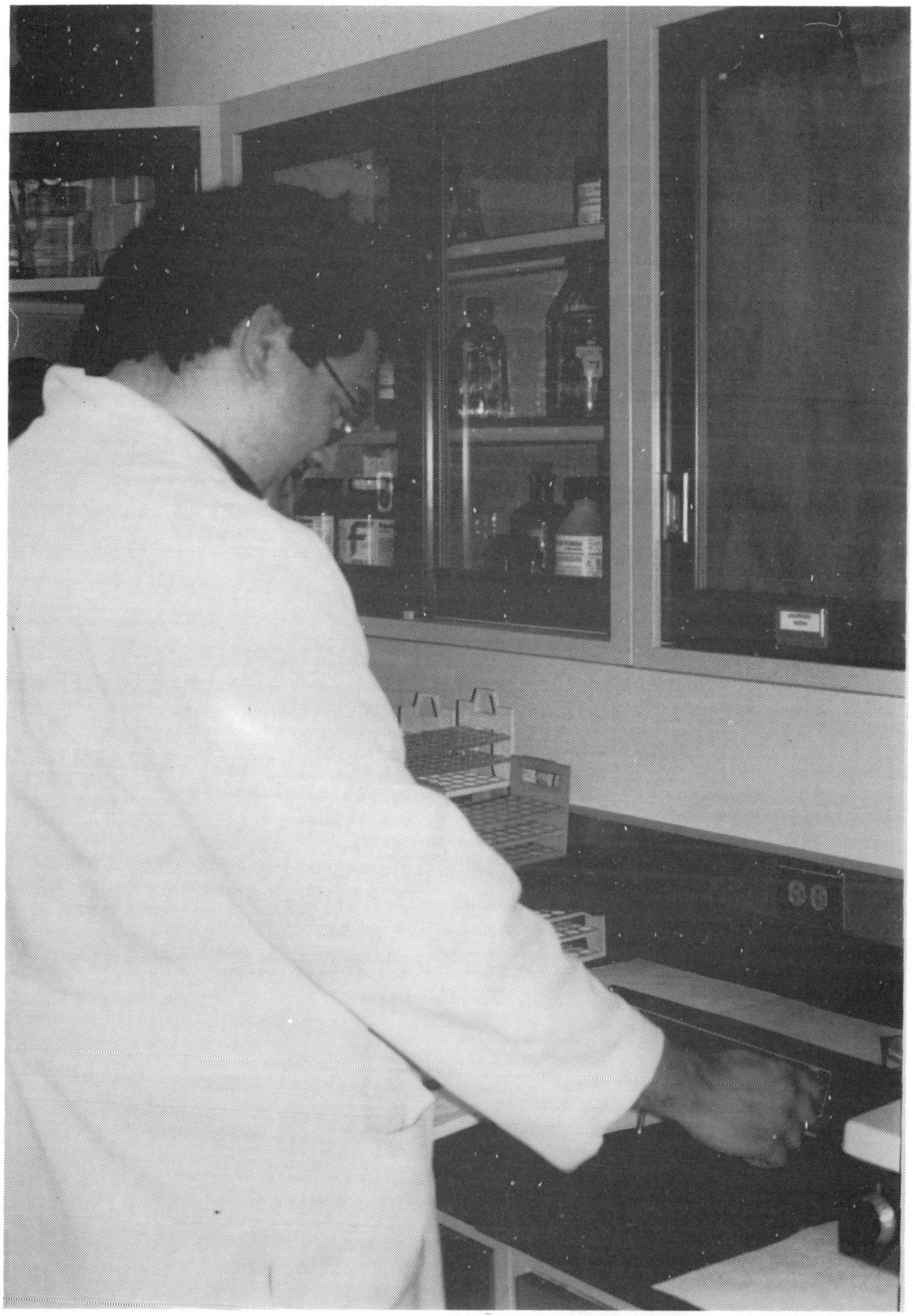

(a)

Figure 15.2. The photographs in Figures 15.2(a)–(k) clearly point out that, in a typical environmental laboratory, glass is everywhere, hence, the need for a strong glassware safety program should be obvious.

(b)

Figure 15.2 (continued)

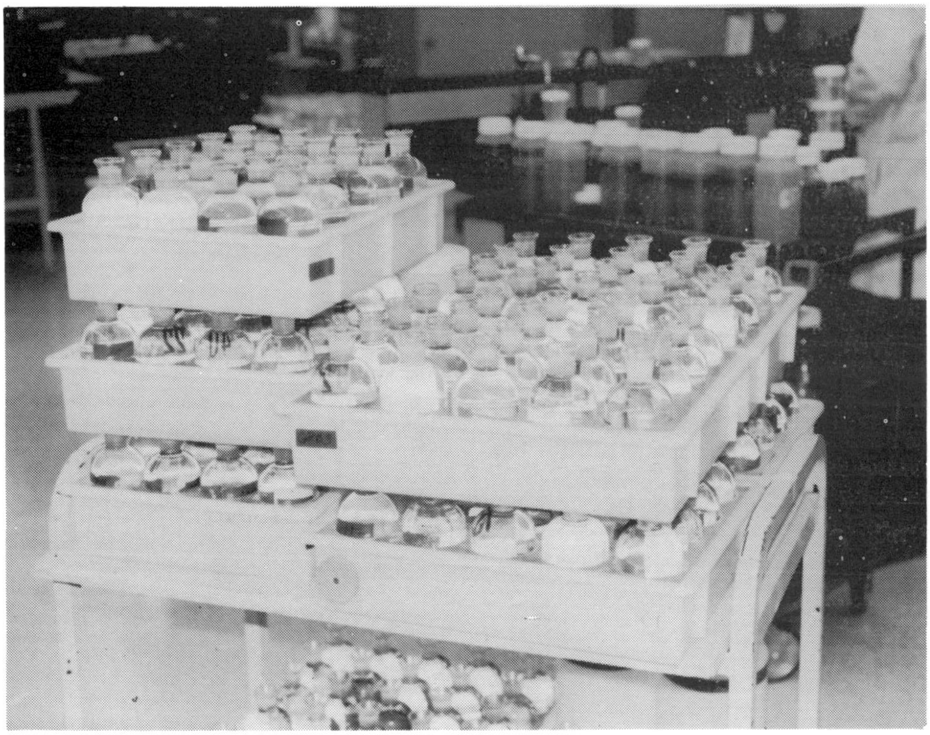

(c)

Figure 15.2 (continued)

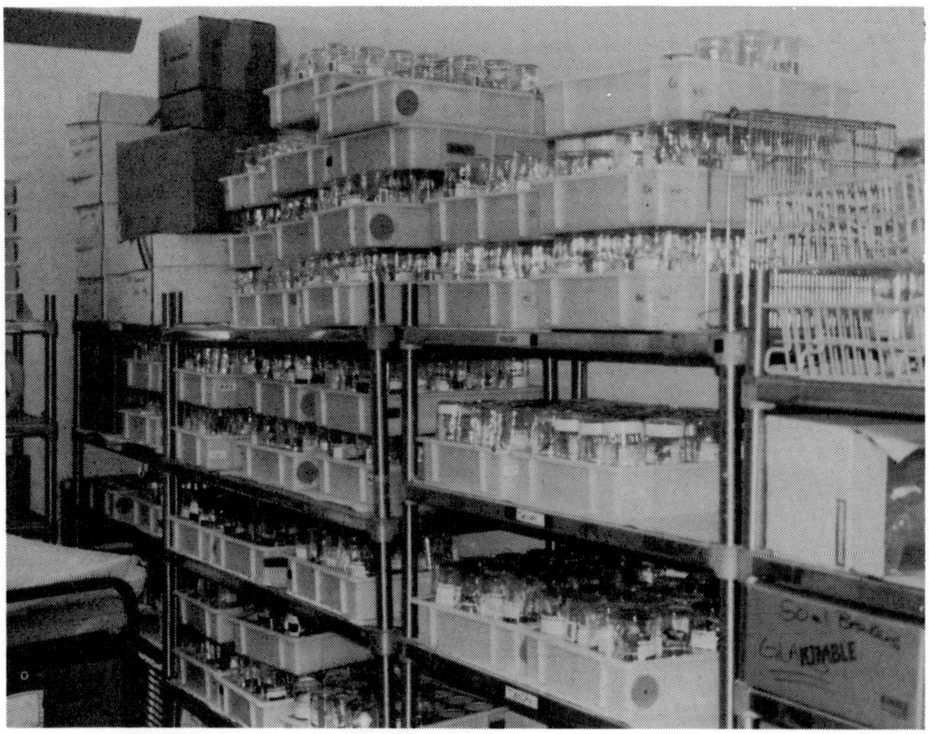

(d)

Figure 15.2 (continued)

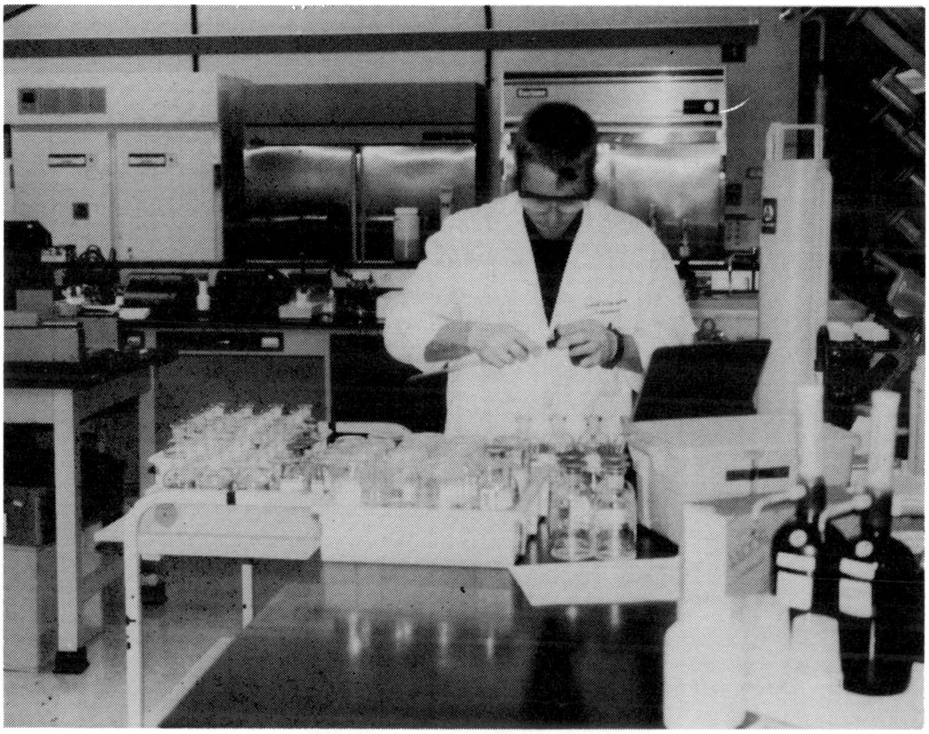

(e)

Figure 15.2 (continued)

(f)

Figure 15.2 (continued)

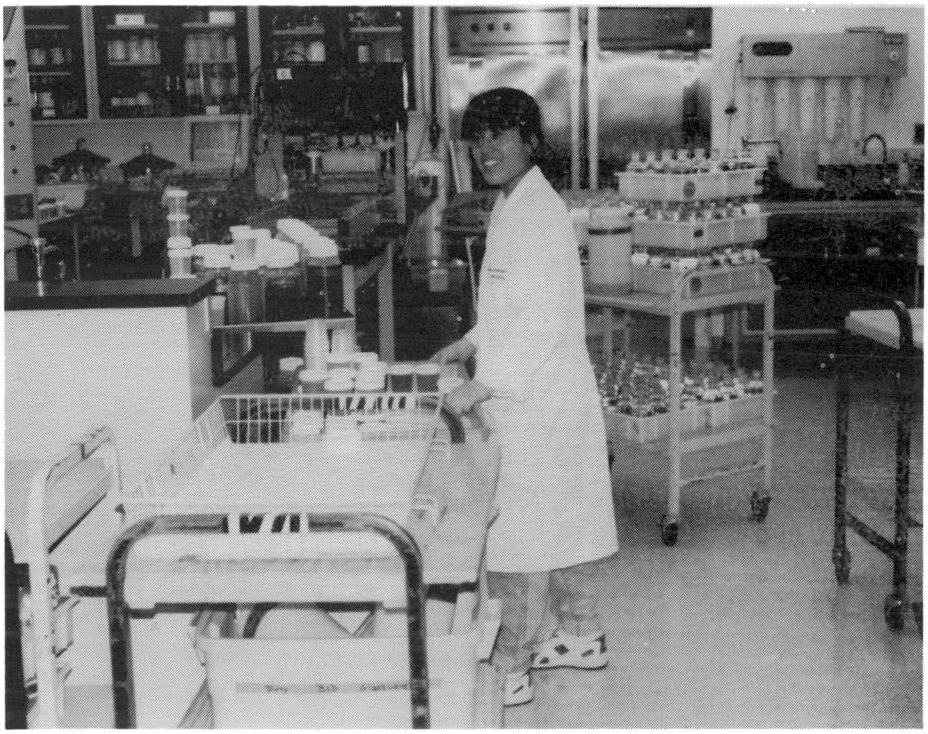

(g)

Figure 15.2 (continued)

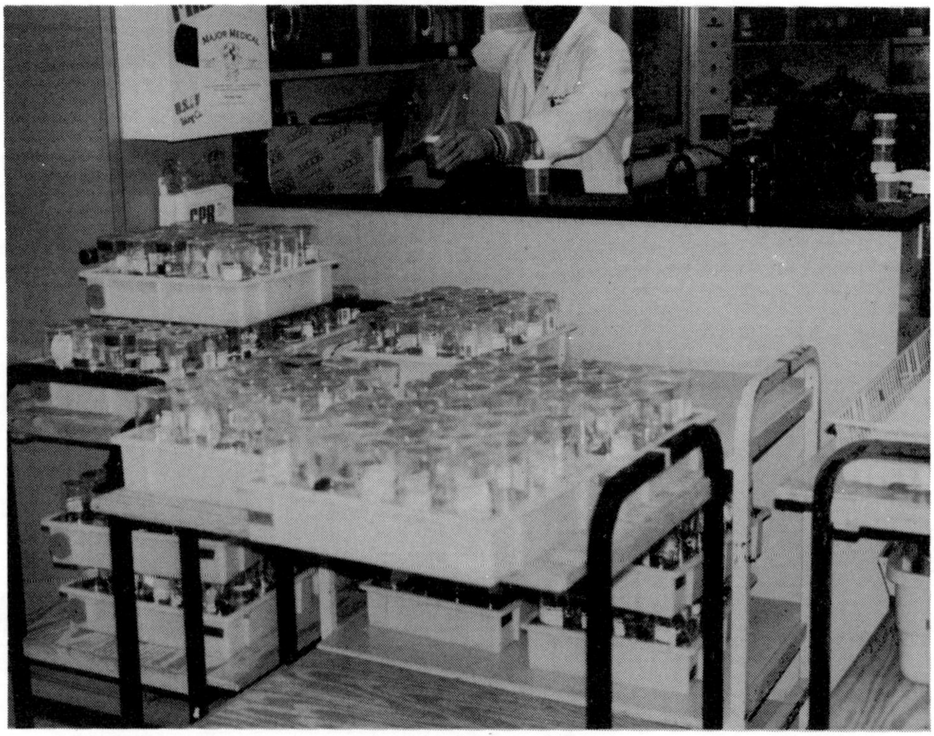

(h)

Figure 15.2 (continued)

(i)

Figure 15.2 (continued)

(j)

Figure 15.2 (continued)

(k)

Figure 15.2 (continued)

Compressed Gas Cylinder—Operation

(1) When received, a cylinder must be checked for damage and for proper contents. An in-house "necklace" label should be placed over the cap of the tank to indicate its contents, that it is full, the date received, and the initials of the person receiving the tank. *Caution*: The Compressed Gas Association warns that color should never be used as the sole means of identifying the contents of a compressed gas container.

(2) Do not attempt to open the main tank valve until the tank has been attached to a pressure regulator or other piece of equipment. Otherwise, the tank may give off a high pressure discharge of gas that cuts like a knife. *Caution*: The Compressed Gas Association warns that the release of high pressure gas from cylinders can be hazardous unless adequate means are provided for reducing the gas pressure to usable levels and for controlling the gas flow. Accordingly, pressure-reducing regulators should always be used when withdrawing the contents of a cylinder, as such devices deliver a constant safe working pressure.

(3) Transport the cylinder in a specially designed gas tank transport cart with the cap in place and with the tank secured to the cart with chains or ties. Never attempt to move the cylinder by rolling it or sliding it along the floor as it may drop to the floor and damage the main cylinder valve.

(4) Store gas cylinders in a dry, well-ventilated area that does not get excessively hot or cold (−29 to 52°C). Do not store incompatible gases in close proximity (see Figure 15.3).

(5) Cylinders must be securely chained or held in place to prevent their falling over (see Figure 15.4). When removing tanks, use the special cart to transport them to the location and immediately chain or secure the tank into position rather than leaving it on the cart.

(6) When attaching the cylinder to the apparatus or pressure regulator, use the proper tool, (wrench) rather than a makeshift device such as pliers. Keep the proper tool chained near the area where the cylinder is used. The apparatus should be securely attached to the main valve so that it will not shift position when lateral pressure is applied to the device. After attaching the device to the main cylinder valve and before opening it, verify that the attached device or pressure regulator is closed off. Then, partially open the main cylinder valve and check for any leaks between the main valve and the attached device. If any leaks are found, close the main valve immediately and seek assistance from the laboratory supervisor. If no leaks are found, check the tank pressure (it should be neither too high nor too low for the particular gas) and verify that the discharge pressure valve is closed or that the final dispensing valve is closed.

Adjust the flow of gas using the pressure adjustment valve (if present) and open the final valve. If any problems are encountered during the use of the compressed gas, turn off the main tank valve first before attempting to resolve the problem. Wear appropriate eye, hand, and body protection when 300 working with the cylinders and use no lubricants on the cylinder or the pressure regulators. It is best to discontinue using the tank before it is completely empty to prevent contaminants from entering the tank during the refilling step at the suppliers. Remove the original necklace label and replace it with another label indicating that the cylinder is ***empty.*** Have the empty tank removed as soon as possible to prevent others from using it and to remove clutter from the laboratory. The tank must be transported back to storage on the cylinder cart with the cylinder cap in place. *Caution*: The Compressed Gas Association warns—do not attempt to force connections that do not fit.

General Safety Considerations for Compressed Gases

The Compressed Gas Association advises or cautions about a number of practices (in addition to those already pointed out above) that should be followed or avoided when handling compressed gas cylinders. These advisories are listed in the following:

(1) Caution: Compressed gases must never be transferred from one container to another except by the gas manufacturer or distributor.

(2) Caution: Do not smoke or permit smoking or open flames in any area where flammable liquids or gases or liquid oxygen are stored, handled, or used, or where they are loaded or unloaded.

(3) Caution: Keep all combustible materials, especially oil or grease, away from oxygen.

(4) Caution: Liquid oxygen, if spilled on clothing or another combustible substance, can pose a serious hazard of fire or explosion due to the rapid chemical reaction between the two substances.

(5) Caution: A flame must never be used for detection of flammable gas leaks.

(6) Caution: Never permit oil, grease, or other combustible materials to come in contact with cylinders, valves, regulators, gases, hose, and fittings used for oxidizing gases such as oxygen and nitrous oxide, which may combine with these substances with explosive violence.

(7) Caution: Poison gases pose serious potential hazards to personnel and therefore require special handling.

(a)

(b)

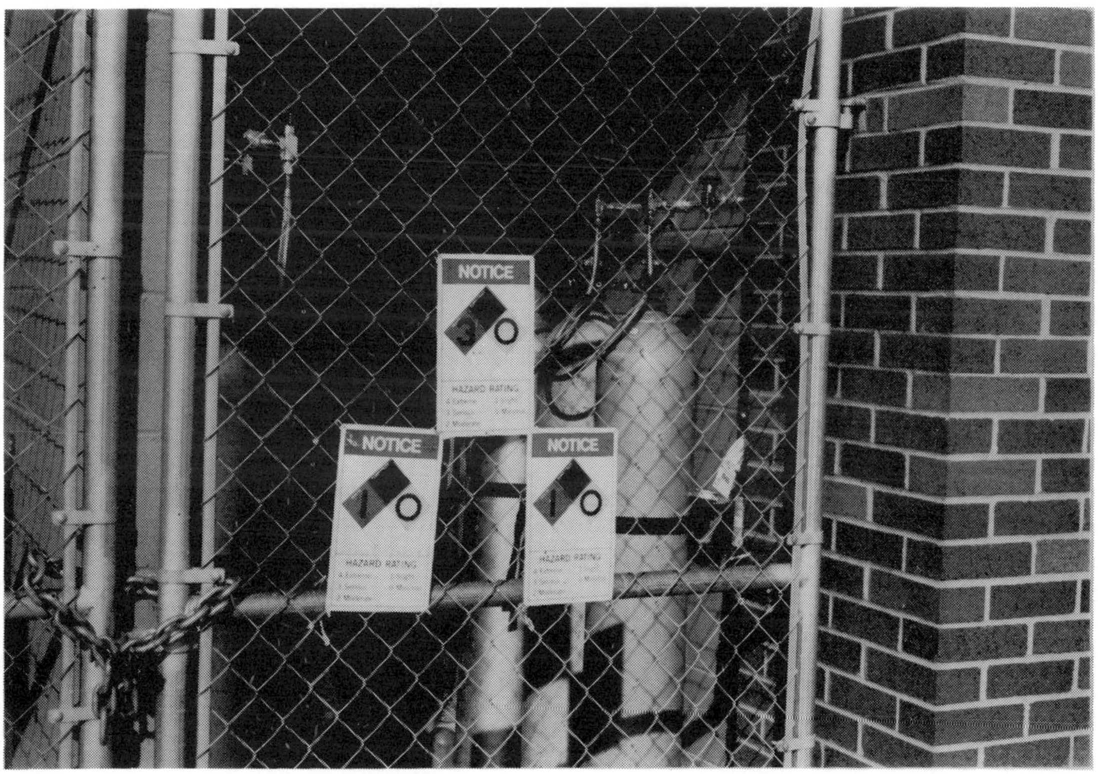

Figure 15.3. Proper storage of compressed gases is important.

(c)

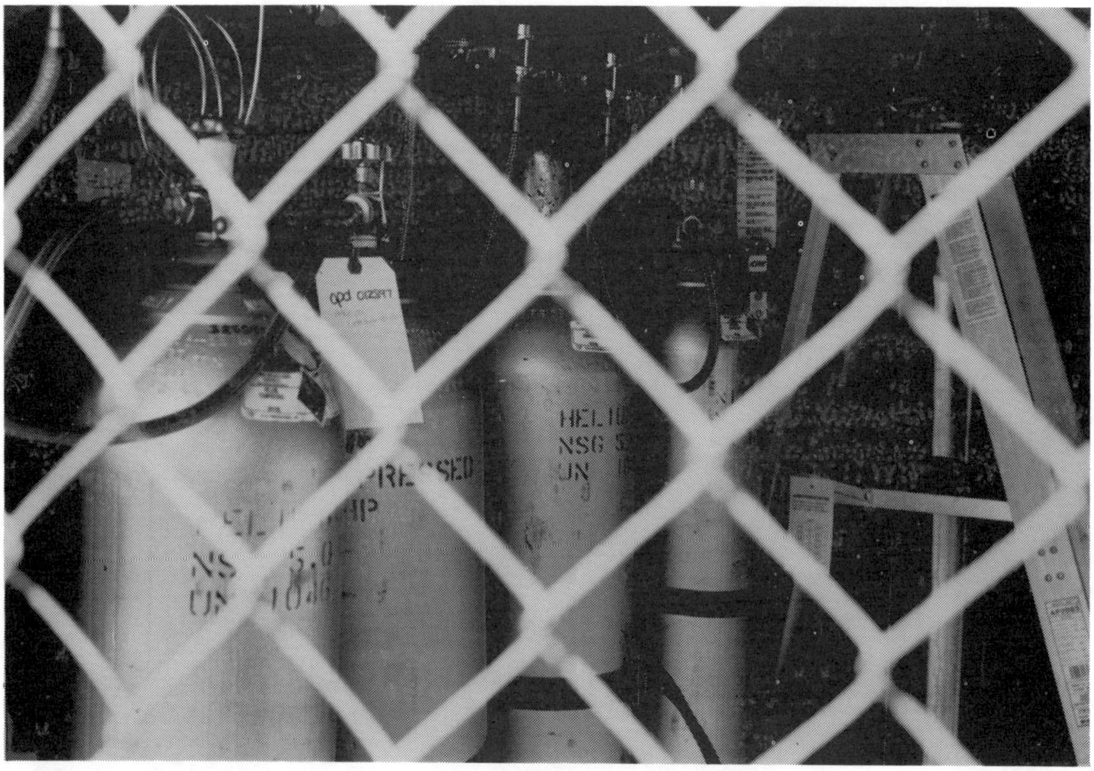

Figure 15.3. *Continued*

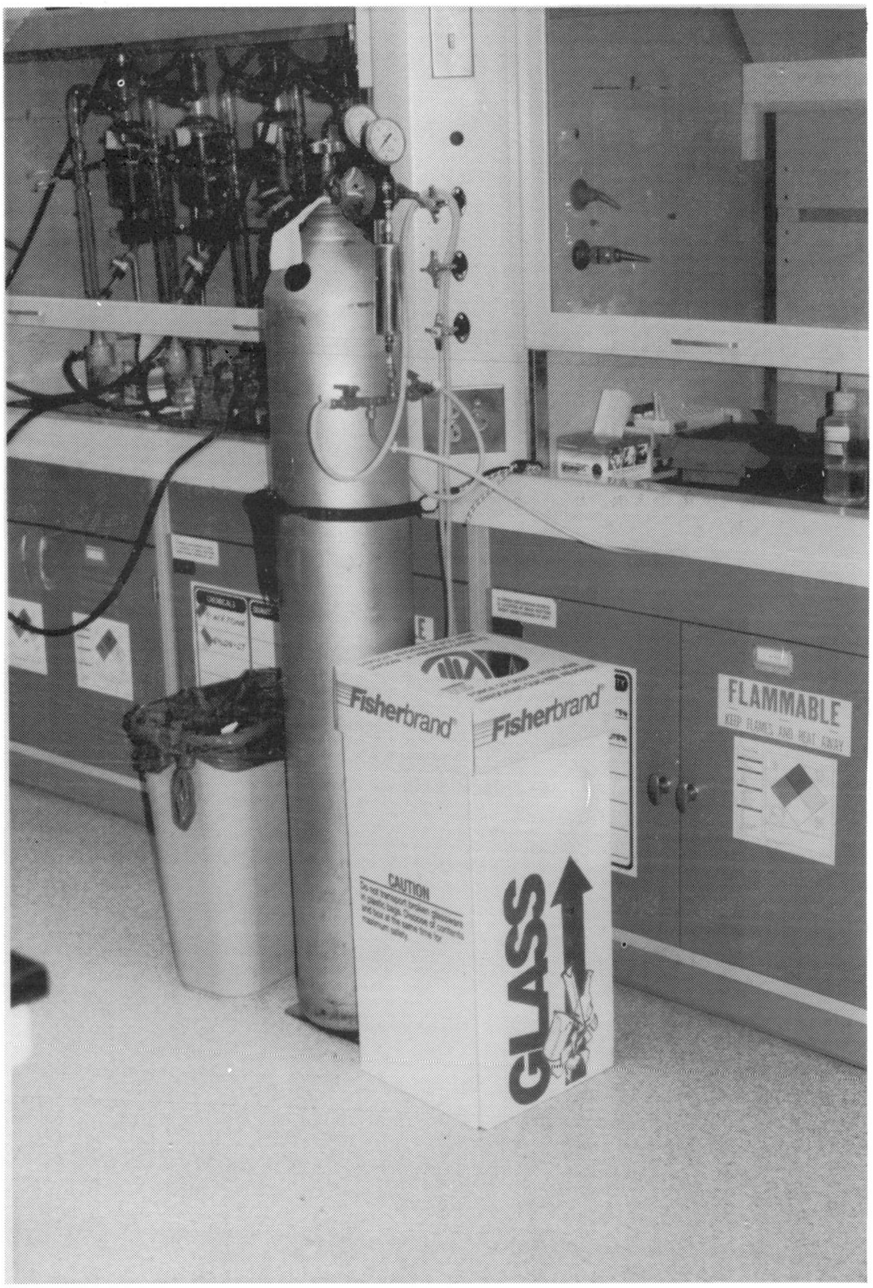

Figure 15.4. Compressed gas must be properly secured in place.

(8) Caution: Corrosive gases attack human tissue and other materials, and special protective clothing and self-contained breathing apparatus must be used by personnel handling these substances.

15.3.8 CENTRIFUGE SAFETY

(1) Before using a centrifuge check to ensure that it is in proper operating condition, review the log book to review any notes from previous users.

(2) Make certain that the proper rotor is being used: that it is rated for the speeds to be used.

(3) Verify that the centrifuge tubes to be used are proper for the rotor and that they can withstand the ***g forces*** that will be generated and the chemicals that will be placed in them.

(4) Stand clear of the rotor as it is accelerating or slowing down. Take care to keep long hair or jewelry away from the spinning rotor.

(5) Never stop or slow a rotor with the hand. Clean up spills inside the centrifuge and especially in the rotor.

(6) Note any unusual noises or operational problems in a log book.

15.3.9 PIPETTING SAFETY

(1) The use of mouth pipetting is strictly prohibited in all laboratories. The danger of ingesting toxic liquids or vapors from the liquids is simply too great to risk such a procedure.

(2) Never point a pipet in the direction of another person.

(3) Broken or chipped glass pipettes are to be disposed of in designated broken glass containers.

15.3.10 FALL PREVENTION

(1) Keep laboratory aisles clear and floors dry.

(2) Use approved ladders or stepstools for reaching upper shelves.

15.4 EMERGENCY RESPONSE: MEDICAL EMERGENCY

Along with providing a written plan, emergency equipment, and worker training for chemical/hazardous materials incidents in the environmental laboratory, it is also important to include medical response planning and training. The lab's medical response plan should be integrated with the emergency action plan or procedures. Figure 15.5 shows a sample medical response guide. This medical response guide has the advantage of being short, to the point, and shown on one sheet of paper. Obviously, this is a user-friendly guide that is designed to facilitate quick and correct response when it is needed.

In addition to having some type of medical response plan it is also important that the environmental laboratory establish emergency first-aid stations. These first-aid stations should be stocked with enough medical supplies to provide general first aid for minor cuts, sprains, and abrasions. In addition to standard first aid supplies, these first-aid stations should contain stretchers, burn kits, ice, emergency eyewash, and fire extinguishing blankets. Emergency medical phone numbers should be posted throughout the laboratory at, near, or on each phone.

As with all types of emergency response, training is a key element. One thing is certain, when hazardous materials/chemicals are involved with any type of emergency response, the responders must be trained to know how to prevent spread of the spill and how to protect themselves from contamination. In addition, key laboratory personnel should be trained in standard first aid—all laboratory personnel should be trained in CPR.

HAMPTON ROADS SANITATION DISTRICT

HRSD

MEDICAL RESPONSE GUIDE

Immediate Procedures

- Remain calm.
- Initiate lifesaving measures if required.
- Call for **Emergency Response:**
- Do not move injured person unless there is danger of further harm.
- Keep injured person warm.

MEDICAL EMERGENCY 911

EMERGENCY RESPONSE PROCEDURES

Clothing on Fire

- Roll person around on floor to smother flame,

  or

  drench with water if safety shower is **immediately available.**
- Obtain medical attention, if necessary.
- Report incident to supervisor.

Chemical Spill on Body

- Flood exposed area with running water from faucet or safety shower for at least 5 minutes.
- Remove contaminated clothing at once.
- Make sure chemical has not accumulated in shoes.
- Obtain medical attention, if necessary.
- Report incident to supervisor.

Biological Spill on Body (Lab Workers)

- Remove contaminated clothing.
- Vigorously wash exposed area with soap and water for 1 minute.
- Obtain medical attention, if necessary.
- Report incident to supervisor.

Hazardous Material Splashed in Eye

- Immediately rinse eyeball and inner surface of eyelid with water continuously for 15 minutes.
- Forcibly hold eye open to ensure effective wash behind eyelids.
- **Obtain medical attention.**
- Report incident to supervisor.

Minor Cuts and Puncture Wounds

- Vigorously wash injury with soap and water for several minutes.
- **Obtain medical attention.**
- Report incident to supervisor.

Figure 15.5. Sample of a medical response guide. Courtesy of HRSD.

PART VII

SAFE WORK PRACTICES

CHAPTER 16

Safe Work Practices: Chemical Handling

> People who work in . . . laboratories are exposed to many kinds of hazards. This can be said of most workplaces; in some, the hazards are well recognized—those of ordinary fire, for example—and the precautions to be taken are obvious. Laboratories, however, involve a greater variety of possible hazards than do most workplaces, and some of those hazards call for precautions not ordinarily encountered elsewhere. In particular, laboratories in which chemicals are used must be prepared to deal with substances known to be hazardous, with the possible hazards of new substances, and with new experiments (National Research Council, 1981, p. 1)

16.1 INTRODUCTION

UNDER the Occupational Safety and Health Administration's (OSHA's) Laboratory Standard, 29 CFR 1910.1450, all laboratory personnel are required to practice prudent chemical handling. This is a sensible requirement especially when one considers that chemical handling operations in the laboratory are likely to be among the riskiest work practices workers perform there.

The risks associated with the physical and health hazards as well as chemical incompatibilities of hazardous chemicals are well known. The sudden, violent, and unforeseen hazards that may result from unsafe chemical handling operations must be guarded against. In order to prevent such mishaps, several safety guidelines should be in place in the laboratory. The purpose of this chapter is to point out several safe work practices and safety guidelines that should be part of the organizational culture of the environmental laboratory.

16.2 SAFE WORK PRACTICES IN THE LABORATORY

(1) Use care and caution when working in the laboratory.
(2) Store chemicals in a safe place where they are not hazardous to personnel or property.
(3) Label all chemical containers, indicating the chemical name and date of preparation and/or container opening.
(4) Check the labels on chemical containers before using to ensure that the proper chemicals are selected for use.
(5) Properly dispose of unlabeled or out-of-date chemicals. *Never* dispose of chemicals in common trash containers. When hazardous waste is to be disposed, contact laboratory supervisor to ensure that proper disposal procedures are followed.
(6) Read and learn the directions for each chemical's use and safety. This information is found on the chemical's Material Safety Data Sheet (MSDS).
(7) Follow directions carefully. *Never* mix chemicals randomly or indiscriminately.
(8) Handle chemicals carefully when pouring or measuring to prevent spillage.
(9) Immediately clean up chemical spills according to the directions on the chemical's MSDS.
(10) Avoid personal contact with chemicals.
(11) Wear correct gloves for the chemical being handled. Refer to the chemical's MSDS if unsure about proper hand protection. *Never* touch chemicals with bare hands.
(12) Ensure that protective gloves are free of cracks, tears, and holes and that gloves fit properly before handling chemicals.
(13) Do not place fingers into mouth, nose, ears, or eyes while handling chemicals.
(14) Wash hands with a disinfectant soap after handling chemicals.

(15) Wash off chemical spills on skin immediately with running water.
(16) Do not smoke or eat in the laboratory while handling chemicals or preforming tests.
(17) Avoid breathing chemical vapors, dusts, or fumes.
(18) Provide positive ventilation to laboratory work areas.
(19) Properly dispose of all broken, chipped, or cracked glassware.
(20) Do not use laboratory glassware as coffee cups or food containers.
(21) Use a suction bulb to pipette chemicals or wastewater. *Never* use mouth to suck up a fluid or chemical in a pipette.
(22) As required by the applicable MSDS, use safety goggles and/or face shield when transferring and measuring chemicals, or whenever there is a potential for chemicals to be splashed in the eyes.
(23) Use tongs or thermal gloves when handling hot utensils.
(24) Use only properly grounded electrical equipment.
(25) Always *add acid to water, not* water to acid.
(26) Use carbon dioxide or dry chemical type fire extinguishers to control laboratory fires.
(27) Ensure that the laboratory is equipped with a properly working emergency eyewash/shower. Laboratory workers should anticipate the need to use emergency eyewash/showers and be familiar with quickest route to each eyewash/shower station in close proximity to their work station.
(28) Ensure that prominent signs warning workers against hot areas such as vents, hot plates, furnaces, water baths, and digestion apparatus are posted in the laboratory.
(29) Inspect acid-neutralizing tanks and basins that service lab sinks regularly and recharge with correct neutralizing agent when needed.
(30) Ensure that emergency phone numbers are posted by all telephones.

16.3 SAFETY GUIDELINES FOR THE LABORATORY

As with safe work practices, general safety guidelines are designed to control chemical hazards in the laboratory. Control of these chemical hazards, of course, should start with engineering controls, followed by administrative controls, and finally by the use of personal protective equipment. Safety guidelines should be included into whenever safety training is conducted for lab workers. Moreover, anytime a standard operating procedure (SOP) is written for a particular laboratory operation it is important to include applicable safety guidelines. In this section several safety guidelines will be outlined. These guidelines should be considered for inclusion into lab's training program and SOPs.

(1) Only the minimum quantities of chemicals consistent with normal laboratory needs and requirements should be maintained.
(2) Before chemicals/hazardous materials are received by laboratory personnel, laboratory management should ensure that workers have been trained on proper handling of chemical, storage requirements, and to ensure that no chemical container is accepted without an adequate identifying label and MSDS. In addition, no container of chemical/hazardous material should be accepted unless it is thoroughly inspected to ensure it is damage and leak free.
(3) Chemicals in storage should be protected to preclude spills, leaks, fire, earthquake damage, etc.
(4) To minimize the risks associated with chemical incompatibility, the laboratory should establish a segregation scheme for chemicals in storage to ensure that accidental breakage, leaks, or other mishaps to chemical containers do not result if they should react with incompatible materials.
(5) Storage practices for flammable liquids (flashpoint less than 100°F) include several guidelines. In general [according to the National Fire Protection Association's (NFPA-45)], proper storage practices include the following:
    - The quantities of flammable chemicals stored in the laboratory should be kept to a minimum.
    - Flammable liquids must be stored away from heat and ignition sources.
    - Small quantities (working amounts) of flammable chemicals may be stored on open shelves.
    - Flammables should not be stored in areas exposed to direct sunlight.
    - Flammable chemicals must be stored in flammable liquid storage cabinets that have been approved by the NFPA. The following safety practices should be observed: (1) store only compatible materials inside the cabinet; (2) follow manufacturers' recommended quantity limits and do not overload cabinets; (3) follow NFPA guidelines for maximum allowable volumes; and (4) do not store paper, cardboard, or other combustible materials in a flammable liquid storage cabinet.
    - Flammable liquids in large quantities should be stored in metal safety cans. The cans should be

used only as recommended by the manufacturer.

(6) Peroxides may be explosive. This is especially the case whenever they are improperly stored or handled. If peroxides are used in the environmental laboratory, the following guidelines should be followed:

- Store peroxide compounds away from heat and light sources.
- Do not use metal containers since some metal oxides can promote the formation of peroxides.
- Always label peroxides with the date they were opened.

(7) Higher chemical exposure risk is associated with laboratory operations such as weighing, pouring, siphoning, mixing, blending, stirring, and shaking. When performing these operations, care and caution are advised.

(8) When handling corrosives (acids and bases) wear appropriate personal protective equipment including clothing, apron, chemical-resistant gloves, and splash goggles/faceshield. Use proper pouring techniques when pouring acids into water and conduct the procedure in a laboratory fume hood. In addition, all dilutions of corrosives must be performed in a laboratory hood.

(9) When handling flammables keep flammable compounds away from ignition source, such as open flame. Moreover, whenever transferring chemical from one container to another or heating chemical in an open container, use a laboratory fume hood.

(10) Whenever chemicals are transported outside the laboratory, the container should be placed in a secondary, non-breakable container. Before moving these containers ensure that caps are tight.

References

Blackman, W.C., (1993). *Basic Hazardous Waste Management.* Boca Raton, FL: Lewis Publishers.

Bretherick, L., (1991). "Overview of Objectives in Design of Safe Laboratories." *Safe Laboratories: Principals and Practices for Design and Remodeling* (eds.) Ashbrook, P.C. & Renfrew, M.M., Chelsea, MI: Lewis Publishers.

Compressed Gas Association, (1990). *Handbook of Compressed Gases*, 3rd ed. New York: Van Nostrand Reinhold.

DiBeradinis, L.J., Baum, J.S., First, M.W., Gatwood, G.T., Groden, E., & Seth, A.K., (1993). *Guidelines for Laboratory Design*, 2nd ed. New York: Wiley.

Grimaldi, J.V., & Simonds, R.H., (1989). *Safety Management.* Homewood, Il: Irwin Press.

Hall, S.K. (1994). *Chemical Safety in the Laboratory.* Boca Raton, FL: CRC Press.

Knutson, G.W.,(1991). "Principles of Ventilation in Chemistry Labs." *Safe Laboratories: Principles and Practices for Design and Remodeling.* (eds.) Ashbrook, P.C. & Renfrew, M.M. Chelsea, MI: Lewis Publishers.

National Academy of Sciences, (1983). *Prudent Practices for Disposal of Chemicals from Laboratories.* Washington, DC: National Academy Press.

National Academy Press, (1989). *Biosafety in the Laboratory: Prudent Practices for Handling and Disposal of Infectious Materials.* Washington, DC: National Academy Press.

National Research Council, (1981). *Prudent Practices for Handling Hazardous Chemicals in Laboratories.* Washington, DC: National Academy Press.

Newton, J., (1987). *A Practical Guide to Emergency Response Planning*. Northbrook, Il: Pudvan Publishing Co.

NIOSH,(1987). *Guide to Industrial Respiratory Protection.* Cincinnati, Ohio: Nation Institute for Occupational Safety & Health.

Occupational Health Guidelines for Chemical Hazards, 3 vols., NIOSH Pub. NO. 81-123, Government Printing Office, Washington, DC., 1981.

Office of the Federal Register, National Archives and Records Administration, *Code of Federal Regulations*, Title 29, Labor, Part 1910.1450, *Occupational Exposures to Hazardous Chemicals in Laboratories*, U.S. Government Printing Office, Washington, DC, July 1, 1992.

Office of the Federal Register, Guidelines on Workplace Safety and Health Program Management, *Federal Register*, Vol. 54, January 26, 1989, p. 3904.

Office of the Federal Register, Notices: Environmental Protection Agency (EPA), Environmental Auditing Policy Statement, *Federal Register*, July 9, 1986.

OSHA, *Managing Worker Safety & Health*, Office of Consultation Programs, U.S. Department of Labor, June, 1992.

Phifer, L.H. (1991). "Safe Laboratory Design in the Small Business." *Safe Laboratories: Principles & Practice for Design and Remodeling.* (eds.) Ashbrook, P.C. & Renfrew, M.M. Chelsea, MI: Lewis Publishers.

Phiffer, R.W. & McTigue, W.R. (1988). *Handbook of Hazardous Waste Management for Small Quantity Generators.* Chelsea, MI: Lewis Publishers.

Saunders, G.T. (1993). *Laboratory Fume Hoods: A User's Manual.* New York: John Wiley & Sons, Inc.

Schaefer, W.P. (1991). "Safe Laboratory Design: A Users Contribution." *Safe Laboratories: Principles and Practices for Design and Remodeling.* (eds.), Ashbrook, P.C. & Renfrew, M.M., Chelsea, MI: Lewis Publishers.

Spellman, F.R. (1996). *Safe Work Practices for Wastewater Treatment Plants.* Lancaster, PA: Technomic Publishing Co.

Stiricoff, R.S. & Walters, D.B. (1995). *Handbook of Laboratory Health & Safety.* 2nd ed., New York: John Wiley & Sons, Inc.

Index